Numerical Solution of
Systems of Nonlinear
Algebraic Equations

ACADEMIC PRESS RAPID MANUSCRIPT REPRODUCTION

Papers Presented at the NSF-CBMS Regional Conference on The Numerical Solution of Nonlinear Algebraic Systems with Applications to Problems in Physics, Engineering and Economics, Held at the University of Pittsburgh, July 10-14, 1972.

NSF-CBMS Regional Conference on "the Numerical Solution of Nonlinear Algebraic Systems with Applications to Problems in Physics, Engineering, + Economics, University of Pittsburgh, 1972

Numerical Solution of Systems of Nonlinear Algebraic Equations

Edited by

George D. Byrne

Departments of Mathematics and
Chemical and Petroleum Engineering
University of Pittsburgh
Pittsburgh, Pennsylvania

AND

Charles A. Hall

Department of Mathematics
University of Pittsburgh
Pittsburgh, Pennsylvania

ACADEMIC PRESS New York and London 1973

ACADEMIC PRESS, INC.
111 Fifth Avenue, New York, New York 10003

United Kingdom Edition published by
ACADEMIC PRESS, INC. (LONDON) LTD.
24/28 Oval Road, London NW1

Library of Congress Cataloging in Publication Data

NSF-CBMS Regional Conference on the Numerical Solution
of Nonlinear Algebraic Systems with Applications to
Problems in Physics, Engineering, and Economics,
University of Pittsburgh, 1972.
Numerical solution of systems of nonlinear algebraic
equations.
1. Equations--Numerical solutions--Congresses.
I. Byrne, George D., ed. II. Hall, Charles A., 1941-
ed. III. United States. National Science Foundation.
IV. Conference Board of the Mathematical Sciences.
V. Title.
QA218.N22 1972 512.9'42 72-88379
ISBN 0-12-148950-7

PRINTED IN THE UNITED STATES OF AMERICA

CONTENTS

CONTRIBUTORS

Mieczyslaw Altman, Department of Mathematics, Louisiana State University, Baton Rouge, Louisiana 70803

William F. Ames, Department of Mechanics and Hydraulics, University of Iowa, Iowa City, Iowa 52240

Kenneth M. Brown, Department of Computer, Information and Control Sciences, University of Minnesota, Minneapolis, Minnesota 55455

Charles G. Broyden, Computing Centre, University of Essex, Wivenhoe Park, Colchester CO4 3SQ, England

George D. Byrne, Departments of Mathematics and Chemical and Petroleum Engineering, University of Pittsburgh, Pittsburgh, Pennsylvania 15213

John E. Dennis, Jr., Computer Science Department, Cornell University, Ithaca, New York 14850

Sven-Åke Gustafson, Institute of Applied Mathematics, Stockholm, Sweden

Gunter H. Meyer, School of Mathematics, Georgia Institute of Technology, Atlanta, Georgia 30332

Samuel Schechter, Stanford Research Institute, Menlo Park, California 94025

Norman M. Steen, Bettis Atomic Power Laboratory, P. O. Box 79, West Mifflin, Pennsylvania 15122

David M. Young, Jr., Center for Numerical Analysis, The University of Texas at Austin, Austin, Texas 78712

PREFACE

This volume contains invited lectures of the NSF-CBMS Regional Conference on the Numerical Solution of Nonlinear Algebraic Systems with Applications to Problems in Physics, Engineering and Economics, which was held July 10-14, 1972. The host for the conference was the Department of Mathematics, University of Pittsburgh and the principal lecturer was Professor Werner C. Rheinboldt of the University of Maryland.

Professor Rheinboldt's lectures will appear in a companion volume, which will be published in the SIAM Regional Conference Series in Applied Mathematics, as required by CBMS. Since his lectures did serve as the main theme of the conference, those contained in this volume serve fairly specific purposes. These purposes include motivating methods for solving nonlinear systems by examining their origins in methods for linear systems, discussing where nonlinear systems arise, reviewing methods for the nonlinear least squares problem, presenting and reviewing specific methods for solving nonlinear problems, and reviewing the contractor theory for nonlinear systems.

The conference committee which was responsible for the arrangements consisted of Professors George D. Andria and Martin J. Marsden, along with the editors.

We gratefully acknowledge the support of the National Science Foundation (Grant GJ-33612); the Conference Board of the Mathematical Sciences for their choice of the University of Pittsburgh as host for the conference; the understanding advice of the Academic Press staff; and the patience and fortitude of our typist, Miss Nancy Brown. We also acknowledge Professor Werner C. Rheinboldt for his thorough preparation of the principal lectures, his personal interest in and advice on the execution of the conference. Finally, we thank the participants in the conference for their enthusiasm and exhiliarating discussions.

Numerical Solution of Systems of Nonlinear Algebraic Equations

NONLINEAR ALGEBRAIC EQUATIONS IN CONTINUUM MECHANICS

W. F. Ames

1. Introduction.

Nonlinear algebraic equations are not ubiquitous in continuum mechanics. However, they do occur regularly and in a variety of forms which are often difficult to analyze. The diversity of forms, ranging from complex polynomials to simultaneous transcendental forms, is discussed here by means of five examples. These correspond to mathematical models of problems in mechanics whose solution depends substantially upon the solution of nonlinear algebraic equations.

The first example arises during the stability analysis of a density stratified gas flow over a liquid. The possible occurrence of periodic free vibrations of a coupled nonlinear system generates the second set of nonlinear equations. The third example is that of diffusion in distinct regions, separated by a moving boundary or interface. Problem four concerns the approximate development, by collocation, of an invariant solution for boundary layer flow of a viscous fluid. The last set of nonlinear equations is obtained when an implicit numerical method is employed to study equations of the form $u_{xx} = \psi(x,t,u,u_x,u_t)$.

2. Polynomials with Complex Coefficients (Sontowski et al., [14]).

Consider two inviscid fluids in a steady state of horizontal streaming (x direction) and superimpose a disturbance upon this state. Assuming the disturbance to be small and neglecting terms of order higher than the first, we obtain six linear perturbation equations in six unknowns as a result of the physical requirements of flow continuity, incompressibility, momentum and interface kinematics. The method of normal modes, discussed by Chandrasekhar [6], is now employed. Solutions are sought whose dependence upon x (horizontal direction), y (normal direction), z (vertical direction) and time t is given by

$$w(z) \exp\left[i\,(k_x x + k_y y + nt)\right]. \tag{1}$$

Upon substitution of (1) into the perturbation equations six equations are obtained which, by a process of elimination, reduce to one equation in one unknown. This equation is

$$\left\{(n+k_x U)\left\{\frac{d^2}{dz^2}-k^2\right\}w - k_x\frac{d^2U}{dz^2}w - \frac{gk^2}{\rho}\frac{d\rho}{dz}\frac{w}{n+k_x U} + \frac{1}{\rho}\frac{d\rho}{dz}\left\{(n+k_x U)\frac{dw}{dz} - k_x\frac{dU}{dz}w\right\}\right\} = 0 \tag{2}$$

where w depends only on z and the product w times $\exp[i(k_x x + k_y y + nt)]$ is the perturbation of the z-component of velocity. The wave number k is defined to equal $(k_x^2 + k_y^2)^{1/2}$. In addition it is required that the quantity $w/(n+k_x U)$ be continuous across the interface and also that

$$\Delta_s\left\{\rho(n+k_xU)\frac{dw}{dz} - \rho k_x\frac{dU}{dz}w\right\}$$

(3)

$$= gk^2\left\{\Delta_s(\rho) - \frac{k^2}{g}T_s\right\}\left\{\frac{w}{n+k_xU}\right\}_s$$

where $\Delta_s(f) = \lim_{\varepsilon\to 0}[f_{z=z_s+\varepsilon} - f_{z=z_s-\varepsilon}]$ and z_s represents the undisturbed interface position.

Application of the above mathematics to the case described in Fig. 1 generates the equations

$$\frac{d^2w}{dz^2} - \beta\frac{dw}{dz} - k^2\left\{1 - \frac{\beta g}{(n+k_xU_a)^2}\right\}w = 0 \qquad z > 0$$

$$d^2w/dz^2 - k^2w = 0 \qquad z < 0,$$

whose general solutions are

$$w = A_a e^{m_+z} + B_a e^{m_-z} \qquad z > 0,$$

$$w = A_b e^{kz} + B_b e^{-kz} \qquad z < 0,$$

where A_a, B_a, A_b and B_b are arbitrary constants and

$$m_\pm = \frac{\beta}{2} \pm \left[\left(\frac{\beta}{2}\right)^2 + k^2\left(1 - \frac{g\beta}{(n+k_xU_a)^2}\right)\right]^{1/2}.$$

Boundary conditions disallow disturbances which

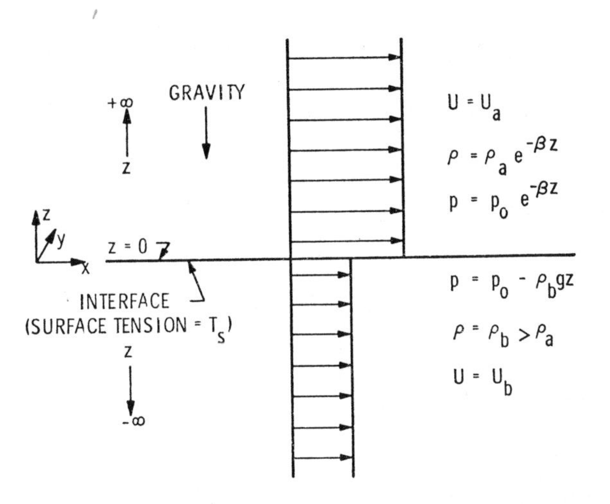

Fig. 1 The assumed stationary state

increase exponentially as the outer bounds of the fluids are approached. Thus

$$w = B_a e^{m-z} \qquad z > 0,$$

$$w = A_b e^{kz} \qquad z < 0,$$

with the requirement that

$$\mathrm{Re}\left[\left(\frac{\beta}{2}\right)^2 + k^2\left(1 - \frac{g\beta}{(n+k_x U_a)^2}\right)\right]^{1/2} \geq \frac{\beta}{2}\,.$$

Continuity of $w/(n+k_x U)$ across the interface leads to the following solution in terms of one arbitrary constant A:

$$w = A(n+k_x U_a)e^{m-z} \qquad z > 0,$$

$$w = A(n+k_x U_b)e^{kz} \qquad z < 0.$$

Substitution into the second interface condition, (3), yields the eigenvalue equation whose dimensionless form is

(4) $\rho_* \beta_k (\nu+k_x \overline{U}_a)^2 - K(\nu+k_* \overline{U}_b)^2 + [(1-\rho_*) + \sigma_k]$

$$= \rho_*(\nu+k_*\overline{U}_a)^2\left(1 - \frac{2\beta_k}{(\nu+k_*\overline{U}_a)^2}\right)^{1/2}$$

in terms of the eigenvalue ν. The additional requirement

$$\text{(5)} \qquad \text{Re}\left(1 - \frac{2\beta_k}{(\nu+k_*\overline{U}_a)^2}\right)^{1/2} \geq \beta_k$$

completes the specification of the eigenvalue problem. According to the definition,

$$\nu = \frac{\left[\left(\frac{\beta}{2}\right)^2 + k^2\right]^{1/4}}{kg^{1/2}} n, k_* = \frac{k_x}{k}, \rho_* = \frac{\rho_a}{\rho_b},$$

$$\beta_k = \frac{\left(\frac{\beta}{2}\right)}{\left[\left(\frac{\beta}{2}\right)^2 + k^2\right]^{1/2}}, \ K = \frac{k}{\left[\left(\frac{\beta}{2}\right)^2 + k^2\right]^{1/2}}, \rho_k = \frac{T_s}{g\rho_b} k^2,$$

$$\overline{U}_a = \frac{\left[\left(\frac{\beta}{2}\right)^2 + k^2\right]^{1/4}}{g^{1/2}} U_a, \quad \overline{U}_b = \frac{\left[\left(\frac{\beta}{2}\right)^2 + k\right]^{1/4}}{g^{1/2}} U_b$$

From the form of the disturbance in (1), it follows that the flow is unstable if and only if any one or more of the eigenvalues ν has a negative imaginary part. For a complete stability analysis the characteristic values of ν must be examined for all values of the vector $k = [k_x, k_y]$.

The nature of the eigenvalues must now be determined. With the substitutions

$$\xi = \nu + k_* \bar{U}_a, \quad \eta = \nu + k_* \bar{U}_b - \frac{\rho_* \beta_k}{K - \rho_* \beta_k}[k_* (\bar{U}_a - \bar{U}_b)], \tag{6}$$

$$\eta_0^2 = \frac{K \rho_* \beta_k}{(K - \rho_* \beta_k)^2}[k_* (\bar{U}_a - \bar{U}_b)]^2 + \frac{(1 - \rho_*) + \sigma_k}{K - \rho_* \beta_k} \tag{7}$$

in (4) and (5) we find

$$\eta^2 + \frac{\rho_*}{K - \rho_* \beta_k} \xi^2 \left(1 - \frac{2\beta_k}{\xi^2}\right)^{1/2} = \eta_0^2 \tag{8}$$

and

$$\mathrm{Re}(1 - 2\beta_k/\xi^2)^{1/2} \geq \beta_k, \tag{9}$$

which, in conjunction with the auxiliary relationship

$$\xi - \eta = (K/(K - \rho_* \beta_k))[k_* (\bar{U}_a - \bar{U}_b)], \tag{10}$$

is equivalent to the eigenvalue problem. Because of (9) this problem is nonalgebraic. Therefore, it is advantageous to construct a parent algebraic system possessing the eigenvalue problem as a subsystem. If all restrictions on $\mathrm{Re}(1-(2\beta_k/\xi^2))^{1/2}$ are removed then (8), together with (10), is equivalent to a fourth degree polynomial in ν and this is taken as the parent system. Let us distinguish two branches of the parent system, calling our eigenvalue problem the principal or P-branch and the remainder of the system, where $\mathrm{Re}(1-(2\beta_k/\xi^2))^{1/2} < \beta_k$, the subsidiary or S-branch. In mathematical form

$$\text{P branch}\begin{cases} \eta^2 + \dfrac{\rho_*}{K-\rho_*\beta_k}\,\xi^2\left(1 - \dfrac{2\beta_k}{\xi^2}\right)^{1/2} = \eta_0^2, \\ \mathrm{Re}\left(1 - \dfrac{2\beta_k}{\xi^2}\right)^{1/2} \geq \beta_k, \end{cases}$$

$$\begin{matrix}\text{S-branch} \\ \text{Part 1} \\ S_1\end{matrix}\begin{cases} \eta^2 + \dfrac{\rho_*}{K-\rho_*\beta_k}\,\xi^2\left(1 - \dfrac{2\beta_k}{\xi^2}\right)^{1/2} = \eta_0^2 \\ 0 < \mathrm{Re}\left(1 - \dfrac{2\beta_k}{\xi^2}\right)^{1/2} < \beta_k, \end{cases}$$

S-branch Part 2 S_2

$$\left\{\begin{aligned} &\eta^2 - \frac{\rho_*}{K-\rho_*\beta_k}\,\xi^2\left(1 - \frac{2\beta_k}{\xi^2}\right)^{1/2} = \eta_0^2, \\ &\mathrm{Re}\left(1 - \frac{2\beta_k}{\xi^2}\right)^{1/2} \geq 0. \end{aligned}\right.$$

The auxiliary equation, (10), must be satisfied simultaneously with each of the above basic branch equations and conditions. The complete solution of the problem depends substantially upon the geometry of the zeros of complex polynomials [11]. A sketch of the real loci of the basic branch equations and the stability zones is shown in Fig. 2.

3. Simultaneous Polynomials with Real Coefficients (Anand [5], Rosenberg [12], Rosenberg and Atkinson [13]).

To study the response of two identical loud speakers enclosed in a cabinet an equivalent coupled spring-mass system (Fig. 3) is proposed as a model. With equal masses and identical cubic nonlinear springs the dimensionless equations of motion are

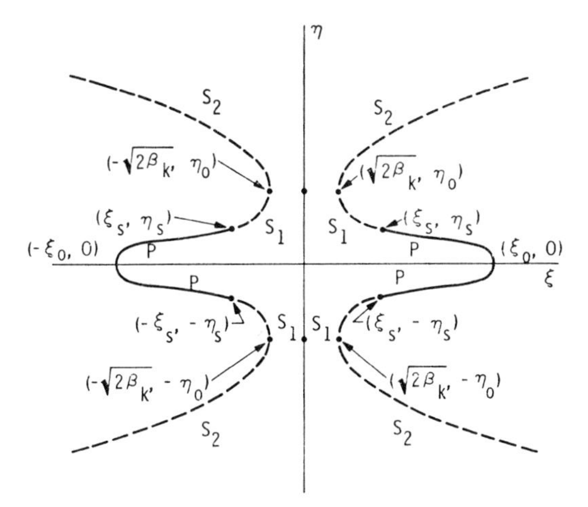

The real locus of the basic branch equations.

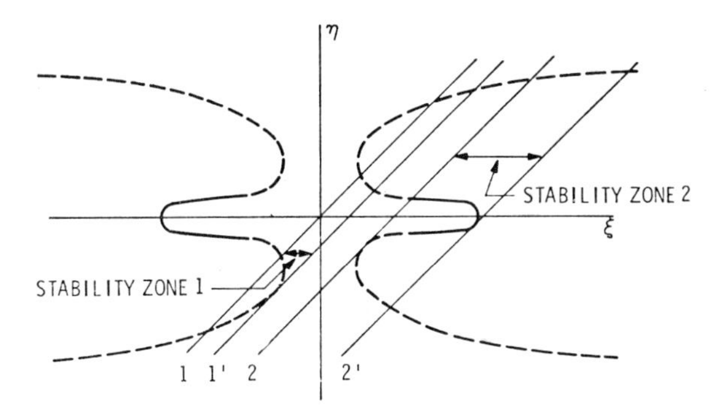

Stability zones in the real (ξ, η)-plane

Fig. 2 Real locus and stability zones

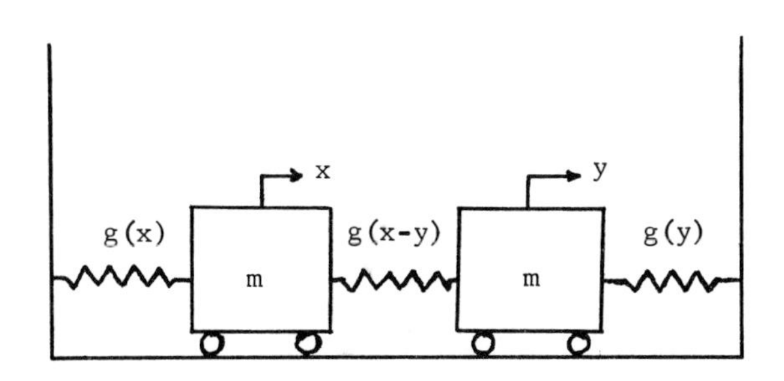

Fig. 3 Nonlinear coupled spring-mass system

$$(11)\qquad \begin{aligned} &\frac{d^2x}{dt^2} + \omega_2^2 x - \omega_1^2 y + \alpha_1 x^3 + \alpha(x-y)^3 = 0 \\ &\frac{d^2y}{dt^2} + \omega_2^2 y - \omega_1^2 x + \alpha_1 y^3 + \alpha(y-x)^3 = 0 \end{aligned}$$

We assume the existence of periodic solutions of (11) in the form

$$(12)\qquad x = A\cos\omega t,\ y = B\cos(\omega t+\beta)$$

and seek to determine relations among A, B, ω and β which permit solutions of this form. Substituting (12) into (11) and equating the coefficients of $\cos\omega t$ and $\sin\omega t$ separately to zero we find

$$(13a)\qquad \left\{\frac{3}{4}(\alpha+\alpha_1)A^3 - \frac{9}{4}\alpha A^2 B\cos\beta + \frac{3}{4}\alpha AB^2(3\cos^2\beta+\sin^2\beta) - \frac{3}{4}\alpha B^3\cos\beta + (\omega_2^2-\omega^2)A - \omega_1^2 B\cos\beta\right\} = 0$$

$$(13b)\qquad \left(\frac{3}{4}\alpha A^2 - \frac{3}{2}\alpha AB\cos\beta + \frac{3}{4}\alpha B^2 + \omega_1^2\right)B\sin\beta = 0$$

$$(13c)\qquad \left\{-\frac{3}{4}\alpha A^3 + \frac{9}{4}\alpha A^2 B\cos\beta - \frac{3}{4}\alpha AB^2(3\cos^2\beta+\sin^2\beta) + \frac{3}{4}(\alpha+\alpha_1)B^3\cos\beta - \omega_1^2 A + (\omega_2^2-\omega^2)B\cos\beta\right\} = 0$$

(13d) $$[-\frac{3}{4}\alpha A^2+\frac{3}{2}\alpha AB\cos\beta-\frac{3}{4}(\alpha+\alpha_1)B^2-(\omega_2^2-\omega^2)]B\sin\beta = 0$$

Periodic free vibrations of the system correspond to the real solutions of (13). A possible but trivial, solution is A = B = 0. The parenthetical expression in (13b) is positive definite for the physically important case $\alpha \geq 0$. For that case non-trivial solutions exist if $\sin\beta = 0$. Thus periodic oscillations are possible only if the masses vibrate either in the same phase or in opposite phases. No other phase relationships are permitted.

Introducing $\sin\beta = 0$ $(0 \leq \beta \leq \pi)$ into (13a) and (13c) we find

(14a) $$(\frac{3}{4}\alpha_1+\frac{3}{2}\alpha)(A^3-B^3)-(\frac{9}{2}\alpha AB+\omega^2-\omega_1^2-\omega_2^2)(A-B) = 0$$

(14b) $$\frac{3}{4}\alpha_1(A^3+B^3) - (\omega^2+\omega_1^2-\omega_2^2)(A+B) = 0.$$

The three pairs of solutions for A and B constitute the possible modes of vibration. Two of these are A = B (symmetric) and A = -B (antisymmetric) while the third is asymmetric.

When g(x) takes another polynomial or rational form, the algebraic problem is considerably altered.

4. Transcendental Equations (Crank [7]).

A number of linear diffusion (heat conduction) problems concern diffusion in two distinct regions separated by a moving boundary or interface. Examples include the progressive freezing of a liquid, tarnishing reactions of a metal and absorption by a liquid of a single component in a mixture of gases. The moving boundary may be marked by a discontinuous change in concentration (second and third example) or by a discontinuity in the gradient of concentration (first example).

We consider the problem of a two step diffusion coefficient D with discontinuities at two concentrations. At the concentration where a discontinuous change in D occurs there is also a discontinuity in the concentration gradient and the way in which this moves must also be determined. The problem is stated mathematically as follows.

Suppose diffusion takes place into a semi-infinite medium and that the surface at $x = 0$ is held at a constant concentration C_1. Also we suppose the concentration to be C_2 (constant) at large distances.

We shall consider a diffusion coefficient D defined by

$$\text{(15)} \quad \begin{aligned} D &= D_1, \quad C_X < c < C_1 \\ D &= D_3, \quad C_Y < c < C_X \\ D &= D_2, \quad C_2 < c < C_Y \end{aligned}$$

Concentrations c_1, c_2, c_3 are associated with the ranges in which the diffusion coefficient has the values D_1, D_2, D_3 respectively. Suppose that at time t the discontinuities in concentration gradient occurring at C_X is at $x = X(t)$ and at C_Y is at $x = Y(t)$. These functions of t have to be determined. At the discontinuity the concentrations must be the same and also the mass of diffusing substance must be conserved.

Region 1: $0 \leq x < X$

$$\frac{\partial c_1}{\partial t} = D_1 \frac{\partial^2 c_1}{\partial x^2}, \qquad 0 < x < X$$

(16)

$$c_1 = C_1, \qquad x = 0$$

Boundary $x = X(t)$:

$$c_1 = c_3 = C_X$$

(17)

$$D_1 \frac{\partial c_1}{\partial x} = D_3 \frac{\partial c_3}{\partial x}$$

Region 2: $X < x < Y$

(18) $$\frac{\partial c_3}{\partial t} = D_3 \frac{\partial^2 c_3}{\partial x^2}$$

Boundary $x = Y(t)$:

$$c_3 = c_2 = C_Y$$

(19)

$$D_3 \frac{\partial c_3}{\partial x} = D_2 \frac{\partial c_2}{\partial x}$$

Region 3: $Y < x < \infty$

$$\frac{\partial c_2}{\partial t} = D_2 \frac{\partial^2 c_2}{\partial x^2} \tag{20}$$

$$c_2 = C_2 \text{ as } x \to \infty$$

The method of solution used herein is due to Neumann. It consists of writing down a particular solution of the differential equations and boundary conditions, (16)-(20) and then seeing what initial condition this solution satisfies. If A, B and E are constants we have the following solutions

$$c_1 = C_1 + A \operatorname{erf}[x/2(D_1 t)^{1/2}], \quad 0 < x < X \tag{21}$$

$$c_2 = C_2 + B \operatorname{erfc}[x/2(D_2 t)^{1/2}], Y < x < \infty \tag{22}$$

$$c_3 = C_X + E\{\operatorname{erf}[x/2(D_3 t)^{1/2}] - \operatorname{erf}[X(t)/2(D_3 t)^{1/2}]\}, \quad X < x < Y \tag{23}$$

The conditions $c_1 = c_3 = C_X,\ x = X$

$c_2 = c_3 = C_Y,\ x = Y$

require that

(24) $$C_X - C_1 = A \operatorname{erf}[X/2(D_1 t)^{1/2}]$$

(25) $$C_Y - C_2 = B \operatorname{erfc}[Y/2(D_2 t)^{1/2}]$$

(26) $$C_Y - C_X = E\{\operatorname{erf}[Y/2(D_3 t)^{1/2}] - \operatorname{erf}[X/2(D_3 t)^{1/2}]\}$$

Since (24)-(26) must be satisfied for all values of t, X and Y must be proportional to $t^{1/2}$, say

(27) $$X = k_1 t^{1/2}, \quad Y = k_2 t^{1/2}$$

where k_1 and k_2 are constants to be determined. By using the two other conditions, (17) and (19), together with (21)-(27), we find two nonlinear algebraic equations,

(28) $$\left\{\frac{D_1^{1/2}(C_X - C_1)e^{-k_1^2/4D_1}}{\operatorname{erf}(k_1/2D_1^{1/2})} - \frac{D_3^{1/2}(C_Y - C_X)e^{-k_1^2/4D_3}}{\operatorname{erf}(k_2/2D_3^{1/2}) - \operatorname{erf}(k_1/2D_3^{1/2})}\right\} = 0$$

$$\left\{\frac{D_2^{1/2}(C_Y - C_2)e^{-k_2^2/4D_2}}{\operatorname{erfc}(k_2/2D_2^{1/2})} + \frac{D_3^{1/2}(C_Y - C_X)e^{-k_2^2/4D_3}}{\operatorname{erf}(k_2/2D_3^{1/2}) - \operatorname{erf}(k_1/2D_3^{1/2})}\right\} = 0 \tag{29}$$

from which to evaluate k_1 and k_2. This system is additionally complicated by the transcendental functions

$$\operatorname{erf} u = 2\pi^{-1/2}\int_0^u e^{-p^2}\,dp, \quad \operatorname{erfc} u = 1 - \operatorname{erf} u$$

appearing in (28) and (29).

This formulation can be easily extended to an m-step diffusion coefficient thereby generating m simultaneous nonlinear equations.

5. Polynomial Equations (Ames [1],[2],[3], Finlayson [8]).

Approximate methods are often employed to develop solutions which are close in some sense to those for a given nonlinear problem. The so-called <u>weighted residual methods</u> constitute one large useful class of such procedures. For a problem

$$(30)\qquad \begin{aligned} &L(u) = f(x) \text{ in } D \\ &B_i(u) = g_i(x), \quad i = 1,2,\ldots,p \text{ on } \partial D \end{aligned}$$

we seek an approximate (trial) solution

$$(31)\qquad \bar{u}(x) = \sum_{j=1}^{n} C_j \phi_j(x) + \phi_0 .$$

Here the $\phi_j(x)$ are chosen to satisfy the boundary conditions (interior method) and the C_j's are undetermined parameters chosen to force the vanishing of a weighted average of the equation residual

$$(32)\qquad \begin{aligned} R[C,\phi] &= f - L(\bar{u}) \\ &= f - L[\phi_0 + \sum_{j=1}^{n} C_j \phi_j(x)]. \end{aligned}$$

The sense in which R is small is that

$$(33)\qquad \int_D w_k R \, dx = 0, \; k = 1,2,\ldots,n.$$

Some of the classical weighted residual methods are as follows:

a) Collocation: $w_k = \delta(p-p_k)$, $k = 1,2,\ldots,n$

b) Galerkin: $w_k = \phi_k$

c) Least squares: $w_k = \partial R/\partial C_k$

d) Subdomain: $w_k(D_k) = 1$, $w_k(D_j) = 0$, $j \neq k$

To illustrate a typical example we consider the approximate calculation of an invariant (similarity) solution [1],[2],[3] for Newtonian flow near a stagnation point. The invariant solution must satisfy the two point boundary value problem

$$\phi''' + 2\phi\phi'' - (\phi')^2 + 1 = 0$$

$$(34) \qquad \xi = 0 : \phi = \phi' = 0$$

$$\xi \to \infty : \phi' = 1$$

Our trial solution is chosen in the form

$$(35) \qquad f(\xi) = \phi_0(\xi) + C_1\phi_1(\xi) + C_2\phi_2(\xi)$$

where the linearily independent ϕ_j's, chosen to satisfy <u>all</u> the boundary conditions, are

$$\phi_0 = \xi - 1 + e^{-\xi}$$

$$\phi_1 = 1 - 2e^{-\xi} + e^{-2\xi} \tag{36}$$

$$\phi_2 = 2 - 3e^{-\xi} + e^{-3\xi}.$$

Upon substituting (35) and (36) into (34) the residual is

$$R = \{\phi_0''' + C_1\phi_1''' + C_2\phi_2''' + 2[\phi_0 + C_1\phi_1 + C_2\phi_2] \bullet$$

$$[\phi_0'' + C_1\phi_1'' + C_2\phi_2''] - [\phi_0' + C_1\phi_1' + C_2\phi_2']^2 + 1\}$$

Upon collocating at two points, $\xi = 1$ and $\xi = 2$, two coupled quadratic equations are obtained. If more terms are desired the number of quadratic equations is increased accordingly.

For non-Newtonian (power law) flows (see [10]) the corresponding equations become

$$(f')^2 - \frac{(2n-1)m+1}{m(n+1)} ff'' = 1 + \frac{1}{m}\frac{d}{dx}[(f'')^n]$$

for Falkner-Skan flow and

$$(f')^2 - \frac{1-2n}{1+n} ff'' = 1 + \frac{1}{c}\frac{d}{dx}[(f'')^n]$$

for Goldstein flows. In both cases $0 < n < 2$.

7. Algebraic Equations in Numerical Analysis (Ames [4]).

Since implicit numerical methods often have certain stability advantages over explicit procedures it is natural to turn to them when stability is threatened.

The equation

$$u_{xx} = \psi(x,t,u,u_x,u_t) \tag{37}$$

typifies certain problems in fluid mechanics. In particular the Burgers' equation

$$u_{xx} = u_t + uu_x$$

is of this type. Additionally the boundary layer equations

$$uu_x + vu_y = \nu u_{yy}, \quad u_x + v_y = 0$$

can be transformed into this type, (37). To accomplish that we define $\psi(x,y)$ by $\psi_y = u$, $-\psi_x = v$

and perform the transformation $(x,y) \to (x,\psi)$, $U = u^2$. The resulting equation is

$$U_{\psi\psi} = \frac{1}{\nu} U^{-1/2} U_x \tag{38}$$

The partial differential equation (37) is properly posed [9] if $\frac{\partial\psi}{\partial u_t} \geq a \geq 0$. With $U_{i,j} = U(i\Delta x, j\Delta t) = U(ih, jk)$ and employing $O(h^2)$ and $O(k)$ approximations for the derivatives we have the implicit algorithm

$$\begin{aligned} \frac{1}{h^2}\delta_x^2 U_{i,j+1} &= \psi[ih, (j+1)k, U_{i,j+1}, \\ & \frac{1}{h}(\mu\delta_x)U_{i,j+1}, (U_{i,j+1} - U_{i,j})/k] \end{aligned} \tag{39}$$

Here $\delta_x U_{i,j+1} = U_{i+1/2,j+1} - U_{i-1/2,j+1}$ and $\mu_x U_{i,j+1} = \frac{1}{2}(U_{i+1/2,j+1} + U_{i-1/2,j+1})$ are central difference and averaging operators, respectively. Determination of the net values on the (j+1) row requires the solution of nonlinear algebraic equations whose complexity depends upon ψ.

REFERENCES

[1] W. F. Ames, Nonlinear Ordinary Differential Equations in Transport Processes, Academic Press, N.Y., 1968.

[2] W. F. Ames, Nonlinear Partial Differential Equations in Engineering, Vol. I, Academic Press, N.Y., 1965.

[3] W. F. Ames, Nonlinear Partial Differential Equations in Engineering, Vol. II, Academic Press, N.Y., 1972.

[4] W. F. Ames, Numerical Methods for Partial Differential Equations, Barnes and Noble, N.Y., 1970.

[5] G. V. Anand, Natural modes of a coupled non-linear system, Internat. J. Non-Linear Mech., 7 (1972) 81-91.

[6] S. Chandrasekhar, Hydrodynamic and Hydromagnetic Stability, Oxford Univ. Press, N.Y., 1961.

[7] J. Crank, The Mathematics of Diffusion, Oxford Univ. Press, N.Y., 1956, Ch. VII.

[8] B. A. Finlayson, The Method of Weighted Residuals and Variational Principles, Academic Press, N.Y., 1972.

[9] A. Friedman, Partial Differential Equations of Parabolic Type, Prentice-Hall, Englewood Cliffs, N.J., 1964.

[10] S-y. Lee and W. F. Ames, Similarity solution for non-Newtonian Fluids, A. I. Ch. E. J., 12 (1966) 700-708.

[11] M. Marden, The Geometry of the Zeros, Math. Surveys, No. 3, Amer. Math. Soc., Providence, R.I., 1949.

[12] R. M. Rosenberg, Normal modes of nonlinear dual-mode systems, J. Appl. Mech., 27 (1960) 263-268.

[13] R. M. Rosenberg and C. P. Atkinson, On the natural modes and their stability in non-linear two-degree-of-freedom systems, J. Appl. Mech., 26 (1959) 377-385.

[14] J. F. Sontowski, B. S. Seidel and W. F. Ames, On stability of the flow of a stratified gas over a liquid, Quart. Appl. Math., 27 (1969) 335-348.

THE NUMERICAL SOLUTION OF QUASILINEAR ELLIPTIC EQUATIONS

Gunter H. Meyer

1. Introduction.

The object of this study is the numerical solution of nonlinear diffusion equations by implicit finite difference methods. It is well recognized that for a large class of problems such solution techniques are slow and cumbersome in comparison to a direct integration of the evolution equation. See, e.g., Price and Varga [10] and Douglas, Jr. and Dupont [1] for a discussion of linear and nonlinear problems, respectively. There are, however, certain nonlinear problems which impose severe stability restrictions on explicit methods but not on implicit methods. Notable among these are conductive heat transfer problems with change of phase, the so-called Stefan problems. Much of our discussion is directed, but certainly not restricted, to the numerical solution of the Stefan problem.

By an implicit method for the diffusion equation we mean the approximation of the time derivative by a suitable backward difference quotient. As is well known the parabolic equation is thus replaced by a sequence of elliptic partial differential equations which we shall write as

$$\sum_{i=1}^{3} \frac{\partial}{\partial x_i} [k_i(u,x,t_n) \frac{\partial u}{\partial x_i}] + \phi(u,x,t_n) = f(x,t_n), \tag{1.1}$$

where $\{t_n\}$ denotes an increasing sequence of time levels, and where the source term $f(x,t_n)$ incorporates the data and (possibly) the history of the evolution system.

It is generally observed (and readily proven for linear systems) that the solution of a parabolic equation by a sequence of elliptic equations is not subject to stability restrictions on the admissible time step. This is an essential feature since in many applications the behavior of a diffusion system has to be modelled over long time periods with (locally) fine spacial resolution. On the other hand, the numerical solution of quasilinear elliptic equations like (1.1) is not yet well understood. From a practical point of view the most important question is the solvability of the nonlinear algebraic equations which result when (1.1) is approximated in a finite dimensional space. This is our central topic. Specifically, we will investigate the algebraic system obtained when the equation (1.1) is solved with finite differences. Our conclusion will be that under mild and (in view of the parabolic origin of the equation) reasonable restrictions on the functions k_i and ϕ the quasilinear problem (1.1) can be solved via a sequence of mildly nonlinear algebraic equations.

Granted that we can solve the nonlinear algebraic system arising from (1.1), there do remain, of course, serious mathematical problems concerning the convergence of the discretized solution to that of the original parabolic equation. While we shall only briefly comment on this aspect in connection with the Stefan problem discussed below, we would like to point out that there exists a considerable body of literature on the solvability and convergence of nonlinear elliptic difference equations. We refer here to the recent account of McAllister [5] where the equations are linearized and then analyzed with fixed point theory. Additional references may be found in this paper. We would like to mention two additional sources. It is well-known that elliptic partial differential equations frequently give rise to monotone operators to which the concept of the degree of a mapping can be applied. Frehse [2] has shown that such techniques carry over to elliptic difference equations and we shall make use of this observation. The second source is the work of Stepleman [13] in which it is already recognized that the discretized quasilinear system largely retains the structure known from linear elliptic difference equations. However, none of the earlier accounts addresses itself to the problem of solving numerically the quasilinear difference equations. This lack of results stands in contrast to the many papers on the numerical solution of mildly nonlinear equations, where only ϕ but not k_i depend on the solution u. For a recent paper and additional sources on this topic we refer to Schryer [12].

2. The Numerical Solution of Some Quasilinear Elliptic Differential Equations.

In order to keep the presentation and notation as simple as possible we shall consider the Dirichlet problem for a one dimensional equation. As will be pointed out repeatedly throughout this section, the numerical algorithm and its properties are independent of the dimension of the elliptic operator.

We shall suppose that the Dirichlet problem is given in divergence form as

$$\frac{d}{dx}[k(u,x)\frac{du}{dx}] - \phi(u,x) = f(x) \tag{2.1}$$

$$u(0) = u(1) = 0.$$

The homogeneous boundary conditions are used in our analysis and may require scaling of the data if non-zero values are given. For multidimensional regions such scaling may be difficult to carry out explicitly but can always be performed in principle provided the boundary and the data are sufficiently smooth. The actual computation does not depend on the boundary values.

The following hypotheses are assumed to apply throughout this paper.

i) The functions k, ϕ, and f are continuous in u and piecewise continuous in x.

ii) There exists a constant k_m such that $0 < k_m \leq k(u,x)$ uniformly in u and x.

iii) For fixed x the function ϕ is monotonely increasing with respect to u.

iv) k and ϕ are locally Lipschitz continuous in u.

Let us comment briefly on these hypotheses, keeping in mind that (2.1) will generally be derived from a parabolic equation of the form

$$\frac{\partial}{\partial x}[k(u,x,t)\frac{\partial u}{\partial x}] - \frac{\partial H}{\partial t}(u,x,t) = f(x,t).$$

We admit piecewise continuity with respect to x in the functions k and ϕ in order to handle those applications involving fixed interfaces, such as heat conduction in a composite slab. The second hypothesis guarantees the ellipticity of the problem. It may be noted that no monotonicity requirement is placed on k. This is important since thermal conductivities may increase or decrease with temperature. The monotonicity of ϕ is a common requirement when discussing mildly nonlinear elliptic equations. It generally can be forced into being through an exponential transformation of the type $u(x,t) = e^{-\omega t}v(x,t)$, $\omega > 0$, applied to the parabolic equation. The Lipschitz continuity for k and ϕ does not come into play in a symmetric manner; indeed, the steeper the slope of ϕ and the flatter that of k relative to that of ϕ, the better are the chances for the applicability of the algorithm to be defined. Finally, we would like to emphasize that the admissible generalization of (2.1) to a multidimensional region is the equation (1.1). Hence we do not require a variational setting and this approach is applicable, for example, to time discretized diffusion in certain anisotropic regions. It remains to be seen to what extent these results hold true for quasilinear equations not in divergence form, i.e., for equations with cross derivatives.

Equation (2.1) will be solved with a standard finite difference scheme. Let $\{0 = x_0 < x_1 < \ldots < x_{N+1} = 1$ be a partition of the unit interval and suppose, for convenience only, that the mesh size $x_i - x_{i-1} = h$ is constant. A central finite difference approximation of (2.1) based on an energy balance at the ith mesh point may be written as

$$\text{(2.2)} \qquad \frac{k(u_{i+1})\dfrac{[u_{i+1}-u_i]}{h} - k(u_i)\dfrac{[u_i-u_{i-1}]}{h}}{h} - \phi(u_i, x_i)$$

$$= f_i \, , \qquad\qquad i = 1,\ldots,N$$

where $u_i = u(x_i)$ and $f_i = f(x_i)$. The boundary conditions are incorporated by setting $u_0 = u_{N+1} = 0$. Should there be interfaces on (0,1) we adjust the mesh spacing so that a mesh point coincides with the interface. The difference equation at this point generally amounts to continuity of flux across the interface. We shall not specifically treat this case.

It is suggestive to write the system (2.2) in matrix form

$$\mathcal{T}(\vec{u}) = A(\vec{u})\vec{u} + \vec{\phi}(\vec{u}) + \vec{f} = 0 \tag{2.3}$$

where $\vec{u} = (u_1,\ldots,u_N)^T$, $\vec{f} = (f_1,\ldots,f_N)^T$, $\vec{\phi}(\vec{u}) = (\phi(u_1,x_1),\ldots,\phi(u_N,x_N))^T$ and where for arbitrary $\vec{y} \varepsilon R^N$ the matrix $A(\vec{y})$ has the entries

$$A(\vec{y})_{ij} = \frac{1}{h^2}\begin{cases} -k(y_i), & \text{if } j = i-1 \\ k(y_i)+k(y_{i+1}), & \text{if } j=i, i=1,\ldots,N \\ -k(y_{i+1}), & \text{if } j = i+1 \\ 0, & \text{otherwise.} \end{cases}$$

The matrix $A(\vec{u})$ is, of course, the discrete

approximation of the differential operator, i.e.

$$A(\vec{u}) \approx \frac{d}{dx}k(u,x)\frac{d}{dx} .$$

It has precisely the structure known for linear elliptic operators, namely: For given $\vec{u} \in R^N$ the matrix $A(\vec{u})$ is a diagonally dominant irreducible M matrix. It may be noted that no use is made of any symmetry properties of the system. Finally, let us observe that the nonlinear system (2.3) is not an M function in the terminology of Ortega and Rheinboldt [9] since the off-diagonal link function

$$\psi_{i+1}(t) \equiv \{-k_i(u_i)u_{i-1} + k_i(u_i)u_i + k_{i+1}(u_{i+1}+t)u_i$$
$$- k_{i+1}(u_{i+1}+t) + h^2[\phi_i(u_i)+f_i]\}$$

for fixed $\vec{u} \in R^N$ is not monotone in t. Similarly, it may be noted that the functions of (2.3) are not diagonally dominant according to the definition of Moré [8] because

$$\{k_i(v_i)v_{i-1} - k_i(u_i)u_{i-1} + k_i(u_i)u_i$$

$$- k_i(v_i)v_i + k_{i+1}(u_{i+1})u_i - k_{i+1}(v_{i+1})v_i$$

$$+ k_{i+1}(v_{i+1})v_{i+1} - k_{i+1}(u_{i+1})u_{i+1}$$

$$+ h^2[\phi(u_i) - \phi(v_i)]\} = 0$$

does not necessarily imply that $|v_i-u_i| \leq \max_{1\leq j\leq N} |v_j-u_j|$.

For ease of typing we shall omit the arrow over vectors in R^N. In order to establish the existence of a unique solution of (2.3) it will be convenient to define on R^N the (discrete $L_2[0,1]$) inner product and norm

$$\langle u,v\rangle = \sum_{i=1}^{N} u_i v_i h, \quad ||u|| = \langle u,u\rangle^{1/2}.$$

Note that this norm is largely independent of h so that, e.g., $||f|| \leq \max_{x\varepsilon[0,1]} |f(x)|$. We can now consider the question of the existence of solutions for (2.3). Important in this context is the uniform invertibility of A(y) for all $y \varepsilon R^N$.

LEMMA 2.1. There exists a constant C_1, which

is independent of h, such that

$$<A(y)u,u> \geq C_1<u,u>$$

for all $y,u \in R^N$.

PROOF. We can write

$2<A(y)u,u> = <[A(y) + A^T(y)]u,u> = \{<[A_m + A_m^T]u,u> + <[(A(y) - A_m) + (A(y) - A_m)^T]u,u>\}$, where $A_m = A(k_m)$ and where the superscript T denotes the transpose. Since $A(y) - A_m$ is itself a diagonally dominant matrix with nonnegative diagonal entries it follows that $<A(y)u,u> \geq <A_m u,u>$. However, A_m is the usual matrix obtained from a central difference approximation for the operator $k_m \frac{d^2}{dx^2}$ (i.e., k_m·Laplacian). It is well known [4, p. 455] that there exists a constant C_1 such that $<A_m u,u> \geq C_1||u||^2$ independently of the mesh size h.

By hypothesis ϕ is locally Lipschitz continuous and increasing with respect to u. Let C_2 be the infimum of all admissible local Lipschitz constants of ϕ. It follows from the monotonicity of ϕ

that

$$[\phi_i(u_i)-\phi_i(0)][u_i-0] \geq C_2 u_i^2.$$

This observation together with Lemma 2.1 is used to establish the existence of a solution for (2.3). We shall use the concept of the invariance of the degree of a mapping under a homotopy [9, p. 156].

THEOREM 2.1. The equation $A(u)u + \phi(u) + f = 0$ has at least one solution and all solutions belong to the set $D = \{u: ||u|| \leq R\}$ where

$$R = \frac{||\phi(0)||+||f||}{C_1+C_2}.$$

PROOF. Let y be arbitrary in R^N and suppose that $||u|| > R$. Then

$$\langle A(y)u,u\rangle + \langle\phi(u),u\rangle + \langle f,u\rangle$$

$$= \langle A(y)u,u\rangle + \langle\phi(u)-\phi(0),u-0\rangle + \langle\phi(0) + f,u\rangle$$

$$\geq \{[C_1+C_2]||u||-||\phi(0)||-||f||\}||u|| > 0.$$

As a consequence no solution of (2.3) can lie outside D. To prove the existence of at least one solution we note from the first part of the proof that the imbedding $H(t,u) = tT(u) + (1-t)u$ satisfies $\langle H(t,u),u \rangle > 0$ for $||u|| > R$ and all $t \in [0,1]$. Hence the degree of $T(u)$ at 0 is one so that there exists at least one solution of $T(u) = 0$.

The use of the degree allows considerable flexibility when dealing with specific problems. If, for example, ϕ is differentiable, $\phi(0) \leq 0$ and $f(x) \leq 0$ then it is readily seen from the nonnegativity of $A^{-1}(y)$ and the positivity of $\phi'(u)$ that all solutions of $A(y)u + \phi(u) + f = 0$ must necessarily belong to the cone $\{u: u_i \geq 0, i = 1,\ldots,N\}$. The properties of solutions for (2.3) come strongly into play when establishing the uniqueness of such solutions. Here we will only assert that $u \in D = \{u: ||u|| \leq R\}$. Since R is independent of h and ϕ is continuous, it is bounded on D. Let C_3 be a constant such that

$$||\phi(u)|| + ||f|| \leq C_3$$

for all $u \in D$. This constant is also independent

of h. We shall now turn to the question of uniqueness of the solution for (2.3) which we cannot resolve without additional growth restrictions on the coefficients. As a building block we need the fact that any solution of (2.3) has bounded gradients.

LEMMA 2.2. Let u be a solution of (2.3). Then

$$\frac{|u_{i+1}-u_i|}{h} \leq \frac{C_3(k_M R+2)}{k_m} .$$

PROOF. The ith equation of (2.3) may be written as

$$k_{i+1}(u_{i+1})\frac{u_{i+1}-u_i}{h} = \Big\{k_i(u_i)\frac{u_i-u_{i-1}}{h}$$

$$+ h\phi_i(u_i) + hf_i\Big\}$$

$$= \sum_{j=1}^{i} [\phi_j(u_j) + f_j]h + k_1(h_1)\frac{u_1-u_0}{h} ,$$

from which follow the inequalities

$$k_1(u_1)\frac{|u_1-u_0|}{h} - C_3 \leq k_i(u_i)\frac{|u_i-u_{i-1}|}{h}$$

$$\leq k_1(u_1)\frac{|u_1-u_0|}{h} + C_3.$$

It remains to show that $\frac{|u_1 - u_0|}{h}$ is bounded. The discretized divergence theorem (summation by parts) yields

$$\sum_{i=1}^{N} k_i(u_i) \left| \frac{u_i - u_{i-1}}{h} \right|^2 h = - \sum_{i=1}^{N} \{\phi_i(u_i) + f_i\} u_i h$$

so that together with the above inequality we can write

$$\frac{1}{k_M} [k_1(u_1) \frac{|u_1 - u_o|}{h} - C_3]$$

$$\leq \frac{1}{k_M} \sum_{i=1}^{N} k_i(u_i) \frac{|u_i - u_{i-1}|}{h} h$$

$$\leq \sum_{i=1}^{N} k_i(u_i) \left| \frac{u_i - u_{i-1}}{h} \right|^2 h \leq C_3 R$$

or $$k_1(u_1) \frac{|u_1 - u_o|}{h} \leq C_3 (R k_M + 1).$$

Substitution into the above inequality leads to the desired result.

It may be noted that Lemma 2.2 is considerably simplified by considering only one space dimension; however, it is not inherently one dimensional and can be carried out analogously in three (or more) dimensions. For ease of reference let us define the constant

$$C_4 = \frac{C_3(Rk_M+2)}{k_m} \quad \text{so that} \quad \frac{|u_{i+1}-u_i|}{h} \le C_4.$$

As mentioned above we shall require that the local Lipschitz constant of $k(u,x)$ is small compared to that of $\phi(u,x)$. To be precise we shall use the following definition.

DEFINITION 2.1. Let $C_2(u)$ and $C_5(u)$ be the local Lipschitz constants of ϕ and k in a neighborhood of u. $C_2(u)$ and $C_5(u)$ are _comparable_ if for given v, w there exists some u on the line between v and w such that $|k(v,x)-k(w,x)| \le C_5(u)|v-w|$, and $|\phi(v,x)-\phi(w,x)| < C_2(u)|v-w|$.

THEOREM 2.2. Suppose that the local Lipschitz constants $C_5(u)$ and $C_2(u)$ are comparable and that $C_5(u) < \frac{2}{C_4}\sqrt{C_2(u)k_m}$. Then the nonlinear system (2.3) has a unique solution.

PROOF. Suppose that there exist two distinct solutions u and v of (2.3). Then $0 = <T(u)-T(v),u-v>$

$= \langle A(u)(u-v), u-v\rangle + \langle [A(u)-A(v)]v, u-v\rangle +$ $\langle \phi(u)-\phi(v), u-v\rangle$. If we use the summation notation for the first two terms and sum by parts we obtain the identity $\langle A(u)(u-v), u-v\rangle + \langle [A(u)-A(v)]v, u-v\rangle$

$$= \sum_{i=1}^{N} k_i(u_i)[\nabla_i(u-v)]^2 h +$$

$$\sum_{i=1}^{N} [k_i(u_i)-k_i(v_i)] \frac{v_i-v_{i-1}}{h} [\nabla_i(u-v)]h,$$ where

$\nabla_i(u-v) \equiv \dfrac{(u_i-v_i)-(u_{i-1}-v_{i-1})}{h}$. Since v is a solution of (2.3) we may apply Lemma 2.2 and the algebraic-geometric mean inequality to estimate the second term by

$$|\langle [A(u) - A(v)]v, u-v\rangle|$$

$$\leq \frac{C_4}{2}\left(\sum_{i=1}^{N} \frac{C_5(w_i)^2}{\alpha_i} |u_i-v_i|)^2 h\right.$$

$$\left. + \sum_{i=1}^{N} \alpha_i |\nabla_i(u-v)|^2 h\right)$$

where w_i lies between u_i and v_i and where the α_i are as yet undetermined positive numbers. Because of the monotonicity of ϕ we can write

$$0 = \langle T(u)-T(v), u-v\rangle \geq \sum^{N} (k(u_i) - \frac{C_4\alpha_i}{2}) \, |\nabla_i(u-v)|^2 h$$

$$+ \sum_{i=1}^{N} (C_2(w_i) - \frac{C_4 C_5^2(w_i)}{2\alpha_i} \, |u_i - v_i|^2 h.$$ If we set

$\alpha_i = \frac{2}{C_4} k_m$ then $C_2(w) - \frac{C_4^2 C_5^2(w)}{4k_m} > 0$ is sufficient to

yield $0 = \langle T(u)-T(v), u-v\rangle > 0$, which is a contradiction.

Some additional comments may help put theorem 2.2 into perspective. The first inclination might be to use a bound like $|\langle [A(u)-A(v)]v, u-v\rangle| \leq CR||u-v||^2$ where C is a constant depending on $C_5(u)$. However, it is readily verified that C is $O(\frac{1}{h})$. Similarly, if it were assumed that $\sum_{i=1}^{N} \frac{|v_i - v_{i-1}|^2}{h} h < \infty$, then C is $O(h^{-1/2})$. Only the fact that v is a solution and therefore has bounded gradients allows the conclusion reached above. The reader may note that the above condition is a monotonicity condition with respect to separate solutions only, rather than uniformly with respect to all vectors u and v. We conjecture that T(u) is both an M function and diagonally dominant at a solution u for sufficiently small h; however, we shall not pursue this topic. We also would like to point out that due to the homogeneous boundary condition the inequality

$$\sum_{i=1}^{N} |u_i - v_i|^2 h \leq \sum_{i=1}^{N} |\nabla_i (u-v)|^2 h$$

is valid. If $\phi(u,x) \equiv \phi(x)$ then $C_2(w_i) \equiv 0$ and the theorem is not immediately applicable. However, in this case we can write $\langle T(u)-T(v), u-v \rangle$

$$\geq \sum_{i=1}^{N} [k(u_i) - \frac{C_4 \alpha_i}{2} - \frac{(C_4 C_5^2)}{2\alpha_i}] \ |\nabla_i (u-v)|^2 h, \text{ where}$$

$$C_5 = \sup_{||u|| \leq R} C_5(u).$$

Thus uniqueness still follows provided C_5 is sufficiently small. Finally, we would like to mention that uniqueness of a classical solution u for (2.1) is readily established if $k(u,x) \equiv k(u)$. Using the Kirchhoff transformation $I(x) = \int_{v(x)}^{u(x)} k(r)dr$, where u and v are two distinct solutions of (2.1), we find that (2.1) is equivalent to

$I'' - [\phi(u) - \phi(v)] = 0$, $I(0) = I(1) = 0$. Nonzero

extrema are now seen to be inconsistent with the positivity of k and the monotonicity of ϕ. One may expect that a similar maximum principle will apply to the discretized equations provided h is sufficiently small. On the other hand, this technique relies strongly on the variational structure of (2.1) which is not shared by equation (1.1). We shall not pursue the maximum principle approach.

Let us now turn to the numerical solution of the nonlinear system (2.3). Under some additional hypotheses we can show that an iteration of the form

$$(2.4) \qquad A(u^n)u^{n+1} + \phi(u^{u+1}) + f = 0, \; u^\circ \text{ arbitrary}$$

converges to the unique solution u* of $T(u) = 0$. This iteration needs some explanation. For given u° the problem $A(u^\circ)u + \phi(u) + f = 0$ describes a discretized mildly nonlinear elliptic difference system which can be solved, at least in principle, with a Gauss-Seidel iteration (see [9, p. 467]). Thus we theoretically can determine the sequence $\{u^n\}$. On the other hand, the iteration (2.4) is not feasible since the equation $A(x)u + \phi(u) + f = 0$ represents a nonlinear system which usually cannot be solved exactly. Instead, only an approximate solution $\hat{u}$ will be obtained which produces a (small) residual $r = A(x)\hat{u} + \phi(\hat{u}) + f$. The convergence theorem for (2.4) will take into account that (2.4) cannot be solved exactly.

THEOREM 2.3. Let $C_2 = \inf_{||u|| \leq R} C_2(u)$, $C_5 = \sup_{||u|| \leq R} C_5(u)$ and suppose

$$C_5 < \frac{2}{C_4} \sqrt{C_2 k_m},$$

Suppose further that $\{\varepsilon_n\}$ is a sequence of positive numbers such that $\varepsilon_n \to 0$. Let u° be arbitrary and let u^{n+1} be the solution of $A(u^n)u^{n+1} + \phi(u^{n+1}) + f = r_{n+1}$, where $||r_{n+1}|| \leq \varepsilon_{n+1}$. Then the sequence u_n converges to the unique solution u^* of $T(u) = 0$.

PROOF. Given u° we can find u^1 in finitely many Gauss-Seidel iterations such that $||r_1|| \leq \varepsilon_1$. By induction we can find $\{u^n\}$ in finitely many steps. Note that each u^n is the solution of a mildly nonlinear elliptic difference system with inhomogeneous term $f-r_n$. It is seen from Lemma 2.2 that $|\nabla_i u^n| \leq C_4 + C_6\varepsilon_n$, where $C_6 = \frac{k_M R+2}{k_m}$. Consider now $\langle T(u^*)-T(u^{u+1}), u^*-u^{u+1}\rangle = \langle -r_{n+1}, u^*-u^{n+1}\rangle$. The estimates in the proof of Theorem 2.2 can be used to yield $\sum_{i=1}^{N} k_i(u^*)|\nabla_i(u^*-u^{n+1})|^2 h$ +

$$\sum_{i=1}^{N} C_2(w^{n+1})\ u_i^*-u_i^{n+1\ 2}h \leq \frac{C_4-C_6\varepsilon_{n+1}}{2}.$$

$$\sum_{i=1}^{N} \frac{C_5^2(w^n)^2}{\alpha_i} |u_i^*-u_i^n|^2 h + \sum_{i=1}^{N} \alpha_i |\nabla_i(u^*-u^{n+1})|^2 h +$$

$$||r_{n+1}||\ ||u^*-u^{n+1}||.$$

Choosing $\alpha_i = \frac{2k_i(u^*)}{C_4+C_6\varepsilon_n}$ we obtain

$$\sum_{i=1}^{N} C_2(w_i^{n+1})|u_i^*-u_i^{n+1}|^2 h \leq \frac{(C_4+C_6\varepsilon_{n+1})^2}{4}.$$

$$\sum_{i=1}^{N} \frac{C_5^{\,2}(w^n)^2}{k_i(u_i^*)} |u_i^*-u_i^n|^2 h + \varepsilon_{n+1}||u^*-u^{n+1}||.$$

With the definition of C_2 and C_5 given above we can write

$$||u^*-u^{n+1}||^2$$

$$\leq \frac{(C_4+C_6\varepsilon_n)^2}{4C_2k_m} C_5^2||u^*-u^n||^2 + \varepsilon_{n+1}||u^*-u^{n+1}||.$$

Since $\varepsilon_n \to 0$ and $\frac{C_4{}^2C_5{}^2}{4C_2k_m} < 1$ we may assume that there exists an integer n_o such that $\frac{(C_4+C_6\varepsilon_n)^2C_5{}^2}{4C_2k_m}$ $\leq \gamma < 1$ for all $n > n_o$. Moreover, since the exact solutions of (2.4) satisfy $||u^n|| \leq R$, we may assume that the perturbed solutions $\{u^n\}$ of our truncated iteration also satisfy $||u^n|| \leq R$ so that finally we obtain the inequality

$$||u^*-u^{n+1}||^2 \leq \gamma||u^*-u^n||^2 + 2R\varepsilon_{n+1}, \quad n \geq n_o,$$

and iteratively,

$$||u^*-u^n||^2 \leq \gamma n-N_0||u^*-u^{N_0}||^2 + \sum_{i=N_o+1}^{n} \gamma^{n-i}\varepsilon_i, \quad n > n_0.$$

Since $\varepsilon_n \to 0$ it follows from the Toeplitz Lemma [9, p, 399] that

$$\lim_{n\to\infty} ||u^*-u^n||^2 = 0.$$

Hence the sequence $\{u_n\}$ of approximate solutions converges to the solution u* of the discretized quasilinear elliptic difference equations.

It has been observed during actual computation that the numerical solution of mildly nonlinear elliptic equations with a Gauss-Seidel method can be very time consuming. Since at each step of the iteration (2.4) such a system has to be solved this problem is magnified for quasilinear equations. It would appear to be advisable to accelerate convergence of this inner iteration through a suitable overrelaxation and we refer to Schechter [11] for a discussion of the choice of relaxation parameters. An alternate approach might be to approximate ϕ by a differentiable function and to try [an underrelaxed] Newton's method. However, since ϕ is generally not convex, no assurance for its convergence can be given.

Finally we would like to point out that the very desirable local condition on the constants of k and ϕ in Theorem 2.2 has been replaced by a global condition. This is troublesome for change of phase problems since ϕ and k are usually discontinuous at the same temperature. It would be preferable to find an iteration for (2.3) which uses only the surjectivity (Theorems 2.1 and 2.3) and the given structure of T(u). We do not yet know how to resolve this question.

3. Two Nonlinear Diffusion Problems.

We shall consider two heat conduction problems which may help put into perspective the hypothesis (i-iv) imposed above on the elliptic equation (2.1).

EXAMPLE 1. Consider the case of heat conduction in a slab of variable conductivity k(u) but otherwise constant thermal properties. A mathematical description of this model is taken as

$$\frac{\partial}{\partial x}\, k(u)\, \frac{\partial u}{\partial x} - \rho c\, \frac{\partial u}{\partial t} = 0 \tag{3.1}$$

$$u(0,t) = \alpha(t), u(1,t) = \beta(t), u(x,0) = u_o(x).$$

For many solids the conductivity can be approximated

locally by a straight line, and we shall write it as

$$k(u) = k_o + k_1 \tan^{-1}\gamma u$$

where, of course, $k_o - |k_1| = k_m > 0$. Replacing $\frac{\partial u}{\partial t}$ by a backward difference quotient $\frac{u_n - u_{n-1}}{\Delta t}$, $n = 1,\ldots,$ we obtain at each time level a nonlinear elliptic system of the form

$$(3.2) \qquad (k(u)u')' - \rho c \frac{u - u_{n-1}}{\Delta t} - 0$$

$$u(0) = \alpha(n\Delta t) \equiv \alpha_h; \; u(1) = \beta(n\Delta t) \equiv \beta_n.$$

In order to provide homogeneous boundary conditions (which are necessary to apply the above theory but not the numerical algorithm) we use the transformation

$$u = v + \alpha_n(1-x) + \beta_n x$$

which, on substitution into (3.2) leads to

$$(k(v)v')' + k'(v)(\beta_n - \alpha_n) - \frac{\rho c}{\Delta t}[v - v_{n-1}] =$$

$$\frac{\partial c}{\partial t}\frac{(\alpha_n - \alpha_{n-1})(1-x) + (\beta_n - \beta_{n-1})x}{\Delta t} \equiv f(x),$$

$$v(0) = v(1) = 0,$$

where $\tilde{k}(v) = k(v+\alpha_n(1-x)+\beta_n x)$. $\tilde{k}'(v)$ is a bounded function of v and will be absorbed into $\phi(v)$, namely

$$\phi(v) = \frac{\rho c}{\Delta t}[v-v_{n-1}] - \tilde{k}'(v)(\beta_n-\alpha_n).$$

We see that for sufficiently small Δt the function $\phi(v)$ is monotone increasing in v. In fact, $\phi'(v) = O(\frac{1}{\Delta t})$. Hence the hypotheses (i-iv) of Section 2 and the proof of Theorem 2.3 will be applicable for sufficiently small Δt.

No computations were carried out with this formulation.

EXAMPLE 2. As a second example we shall consider a heat transfer problem with change of phase. Suppose a heated octagonal duct (meant to approximate a pipe) is buried in moisture laden frozen soil (permafrost). If z is the downward and x the horizontal coordinate then with reference to Fig. 2 the heat flow around the duct can be described by the following equations.

Energy balance in the soil:	$\nabla \cdot k(u) \nabla u - \frac{\partial H}{\partial t} = 0$
Temperature of the duct:	$u = T$ (constant)
Constant temperature at sufficient depth:	$u = T_Z, \ z = z$
Specified periodic surface temperature:	$u = \alpha(t), \ z = 0$
No horizontal heat flow across the two lateral boundaries:	$\frac{\partial u}{\partial x} = 0, \ x = 0, \ x = X$
Specified initial temperature distribution:	$u(0,x,z) = u_o(x,z)$.

Here ρ is the (constant) density of the soil, $k(u)$ its conductivity and $H(u)$ its enthalpy. Strictly speaking both enthalpy and conductivity have jump discontinuities at the freezing temperature T_o. However, in practice freezing occurs over a nonzero temperature range, denoted by $[T_o-\varepsilon, T_o+\varepsilon]$, and it is reasonable to approximate the jump in H by a linear ramp as shown in Fig. 2. A similar, but decreasing, piecewise linear function is used to describe $k(u)$. A complete description of this model may be found in the engineering literature [3], while its mathematical and numerical properties were examined in [6]. In particular, convergence of the discrete solution to the unique weak solution of the continuous Stefan problem was established.

After the discretization described in Section 2 was carried out the above equations lead to a matrix equation of the form

$$A(u)u + \phi(u) + f = 0.$$

However, its solution was computed, not from a sequence of mildly nonlinear equations, but directly from a Gauss-Seidel iteration

$$M(u^{n+1},u^{n})u^{n+1} + \phi(u^{n+1}) = N(u^{n})u^{n} - f,$$

where $M(u^{n+1},u^{n}) = D(u^{n+1},u^{n}) - L(u^{n+1})$ and $N(u^{n}) = U(u^{n})$ are obtained from a regular splitting of the M matrix $A(u)$.

While our convergence results do not apply to this iteration- the algorithm worked well but was slow. Overrelaxation at all points was not always successful; it led to cycling and loss of convergence. It appeared that, because of the steepness of ϕ in the phase transition zone, only a very slight over-relaxation was possible which did not materially affect convergence. However, it was found that overrelaxing at mesh points away from the phase transition zone and no relaxing near this zone could speed up the computation by a factor of 4. The optimum relaxation factor was determined by trial and error.

Fig. 1 shows a computer plot of the phase transition isotherms around a noninsulated duct in permafrost. Fig. 2 shows schematically the enthalpy employed. The technical data listed in [3] were used for this computation.

A final comment. The fall isotherm (#1 in Fig. 1) indicated that the free interface between the liquid and solid phases may be discontinuous. A direct tracking of this interface therefore does not appear feasible. The same is not true for a one dimensional Stefan problem; for such a problem the algorithm outlined here is likely to be very inefficient compared to the more direct methods which explicitly use the free interface. For additional comments on one dimensional Stefan problems see, e.g., [7].

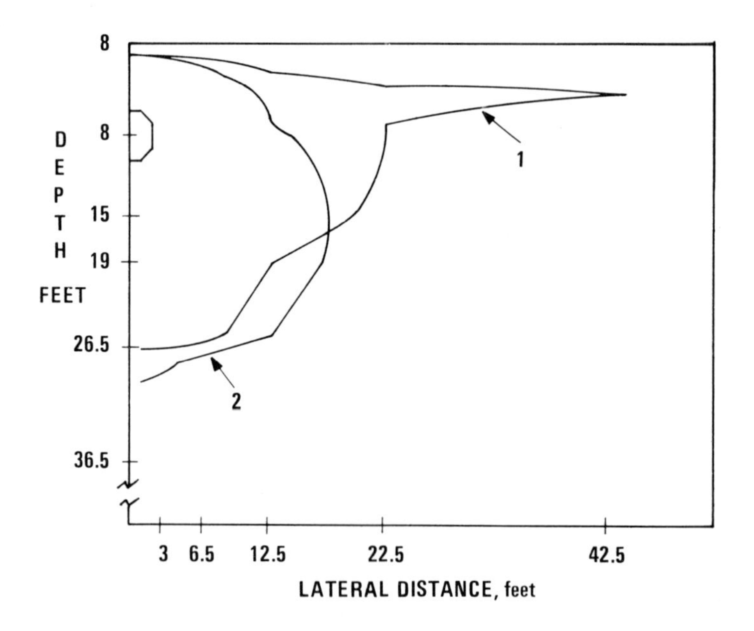

FIGURE 1. Melting isotherms around a non-insulated pipeline in permafrost. Contour 1: t = 182 days (fall); contour 2: t = 365 days (spring).

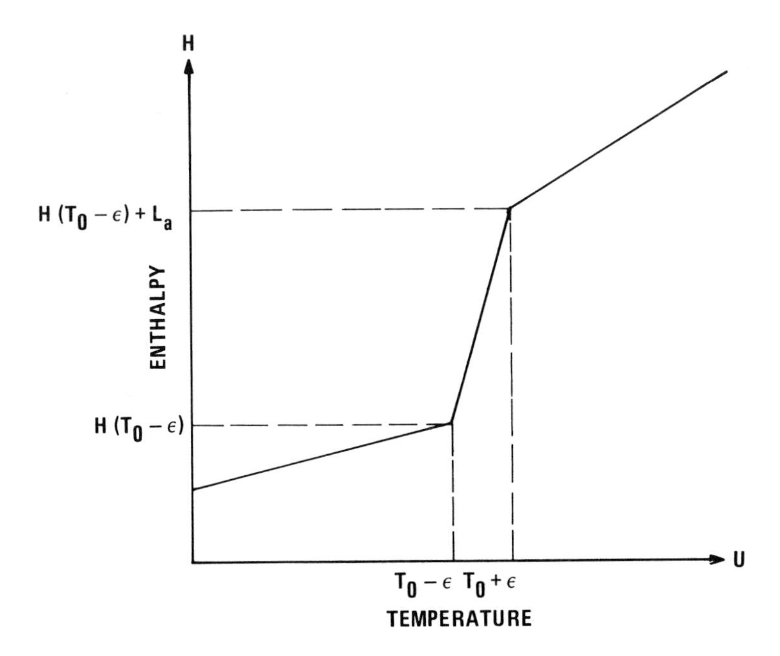

FIGURE 2. Continuous enthalpy approximation for two-phase systems.

REFERENCES

[1] J. Douglas, Jr. and T. Dupont, Galerkin methods for parabolic equations, SIAM J. Numer. Anal. 7 (1970) 575-626.

[2] J. Frehse, Existenz und Konvergenz von Losungen nichtlinearer elliptischer Differenzengleichungen unter Dirichlet-Randbedingungen, Math. Z. 109 (1969) 311-343.

[3] E. J. Couch, Jr., H. H. Keller and J. W. Watts, Permafrost thawing around producing oil wells, J. Can. Pet. Tech., April-June 1970.

[4] E. Isaacson and H. B. Keller, Analysis of Numerical Methods, J. Wiley, 1966.

[5] G. T. McAllister, The Dirichlet problem for a class of elliptic difference equations, Math. Comp. 25 (1971) 655-673.

[6] G. H. Meyer, Multidimensional Stefan problems, SIAM J. Numer. Anal., to appear.

[7] G. H. Meyer, A numerical method for two phase Stefan problems, SIAM J. Numer. Anal. 8 (1971) 555-569.

[8] J. J. More, Nonlinear generalizations of matrix diagonal dominance with applications to Gauss-Seidel iterations, SIAM J. Numer. Anal. 9 (1972) 357-378.

[9] J. M. Ortega and W. C. Rheinboldt, Iterative Solution of Nonlinear Equations in Several Variables, Academic Press, N. Y., 1970.

[10] H. S. Price and R. S. Varga, Error bounds for semidiscrete Galerkin approximations of parabolic problems with applications to petroleum reservoir mechanics, SIAM-AMS Proc. II, 74-94, Amer. Math. Soc., Providence, R. I., 1970.

[11] S. Schechter, On the choice of relaxation parameters for nonlinear problems, this volume.

[12] N. L. Schryer, Solution of monotone nonlinear elliptic boundary value problems, Numer. Math. 18 (1972) 336-344.

[13] R. S. Stepleman, Difference analogues of quasilinear elliptic Dirichlet problems with mixed derivatives, Math. Comp. 25 (1971) 257-269.

NONLINEAR SYSTEMS IN SEMI-INFINITE PROGRAMMING

Sven-Åke Gustafson*

1. Introduction.

Let K be a closed bounded subset of R^n, the n-dimensional Euclidean space and let G be a convex continuous function defined on K. We shall assume that the supporting planes of K are given explicitly. That is, we know a set S of linear inequalities such that

$$\sum_{r=1}^{n} y_r u_r(x) \geq u_{n+1}(x), \qquad x \in S$$

if and only if $y \in K$. Here S is a compact subset of R^k $(k < \infty)$. We shall study the computational solution of the problem (later referred to as task D):

$$\text{Determine } \min G(y), \ y \in K$$

*This research was supported by NSF under Grant GK-31833 and the Institute of Applied Mathematics, Stockholm. The author was Visiting Professor, SUPA, Carnegie-Mellon University, June-July 1972.

when certain general assumptions are placed on u_r, $r = 1,2,\ldots,n+1$. We shall not require that we know a priori a finite set of relations, which can replace S. We call D a semi-infinite program. Such problems occur in numerical analysis (evaluation of linear functionals, approximation of functions, estimation of error bounds), statistics and in technical problems such as the study of the propagation of water and air pollution, cf. [8].

In §2 we illustrate how semi-infinite programs arise from commonly used diffusion models in air pollution abatement studies. For further examples see [7].

In §3 we demonstrate how the optimal solution of a semi-infinite program can be constructed from the solution of a nonlinear system with a finite number of variables and equations. We remark here that the number of unknowns generally is not given from the outset but must be determined from the solution of approximations to the original task D. The results given in §3 were previously obtained in [8] but the derivations here are much simpler. A result in [11] enables us to replace a convex semi-infinite program with an equivalent linear semi-infinite program. We could apply the theory of [8] and [9] to the latter, but we present instead a simplified treatment using the separation theorem for convex sets in R^n and the Carathéodory theorem as major tools. The present treatment is much in the spirit of [10].

In §4 we discuss the numerical treatment of the system constructed in §3. We show that an arbitrarily good approximate solution of the convex semi-infinite program can be constructed using linear programming. We also discuss the determination of the form of a nonlinear system whose solutions are used to treat the original task.

2. An illustrative example.

The following problem often occurs in air pollution abatement. In a given (two-dimensional) control area S one wants to guarantee a minimum air-quality. This goal is assumed to be met if the concentration of a certain pollutant, e.g., SO_2 or CO, averaged over the year is below a given standard, specified by the function ϕ defined over S. (The more complicated cases when standards for several pollutants have to be met simultaneously and/or both short and long time standards are established, can be treated within the same theoretical frame-work. In the present discussion we can therefore confine ourselves to the simpler case indicated).

The observed concentration can be split up into two parts:

A) pollution from sources whose contributions are supposed to be regulated and,

B) pollution from uncontrollable sources.

To the latter category may belong factories outside the area, contributions from transportation, heating, etc.

We assume now that the controllable pollution

comes from n sources whose contributions to the average concentration is described by the n functions $u_1, u_2, \ldots, u_n$ and that the uncontrollable pollution is given by u_0. Then the condition

$$(1) \qquad \sum_{r=1}^{n} u_r(x) + u_0(x) \leq \phi(x), \qquad x \in S$$

should be met. We note that (1) is a system of an infinite number of inequalities. Here x is the position vector of an arbitrary point in S. The actual determination of the functions $u_0, u_1, \ldots, u_n$ from field data is a difficult task which is not completed, cf. [5], [7], [12] and [13].

If the constraint (1) is violated one has to reduce the contributions from the sources associated with $u_1, u_2, \ldots, u_n$. Let E_r be the factor with which the contribution from source r is reduced. Then the condition takes the form

$$\sum_{r=1}^{n} (1-E_r)u_r(x) + u_0(x) \leq \phi(x), \quad x \in S$$

$$0 \leq E_r \leq 1.$$

This system is consistent if and only if $u_0(x) \leq \phi(x)$, $x \in S$.

If we have to select $E_r \neq 0$ a cost is incurred. (It might be necessary to change production schedules, install purifying equipment, etc.). It seems to be rational to minimize the combined control cost which we assume to be given by a function G. We arrive at the problem:

$$\text{Compute } \min_{E} G(E_1, E_2, \ldots, E_n)$$

subject to

$$\sum_{r=1}^{n} E_r u_r(x) \geq \sum_{r=0}^{n} u_r(x) - \phi(x) \qquad x \in S \tag{2}$$

$$0 \leq E_r \leq 1, \qquad r = 1,2,\ldots,n.$$

We shall assume that G is continuous and convex. (In particular it can be linear).

It is possible to prove that there are q points $x^1, x^2, \ldots, x^q$, $q \leq n$ such that equality occurs when an optimal solution is substituted in the inequalities (2). We shall show that $x^1, x^2, \ldots, x^q$

can be constructed from the solutions of a certain system of nonlinear equations, cf. [8].

The task discussed in this section is a semi-infinite program, see [2] and [3]. This type of problem appears in various fields such as the study of the propagation of water- and air pollution, theory of signals, approximation of functions, statistics and numerical analysis [6], [8], [9].

In the following sections we shall give a general treatment of the numerical solutions of semi-infinite programs.

3. General results on semi-infinite programs.

We introduce a few notations and assumptions which will be used in the sequel.

Let S be a compact subset of R^k, the k-dimensional Euclidean space ($k < \infty$) and let $u_1, u_2, \ldots, u_{n+1}$ be continuous functions defined on S. Let F be a positive constant and denote by K the set

$$K=\{y \mid \sum_{r=1}^{n} y_r u_r(x) \geq u_{n+1}(x);\ x \varepsilon S;\ |y_r| \leq F,\ r=1,2,\ldots,n\}.$$

K is a compact subset of R^n, if it is non-empty. It will be called the <u>constraint set</u> of the problem

Now let G be a given function, which is convex and has continuous derivatives of the first order on K.

We consider the problem:

$$\text{Compute } \min G(y), \quad y \in K$$

We shall henceforth refer to this as task D, which is a semi-infinite program. It always has a solution under the assumptions made here and if G is strictly convex then the solution is unique. We note that this problem subsumes the task discussed in the preceding section. For other applications, see [8].

LEMMA 1. Let the conditions made above on $u_1, u_2, \ldots, u_{n+1}$ and G prevail and assume that K is nonempty. Denote by y^* an optimal solution. Then y is an optimal solution of the program $\min_{y \in K} G(y)$ if and only if y is also an optimal solution of the linear semi-infinite program

$$\min_{y \in K} \sum_{r=1}^{n} y_r \left(\frac{\partial G}{\partial y_r} \right)_{y=y^*} .$$

The proof can be found in [11].

Hence in our theoretical analysis we need only to consider the linear case.

Let μ be a given vector. Then we define problem $(\tilde{D})$:

Compute $\inf \sum_{r=1}^{n} y_r \mu_r$

subject to $\sum_{r=1}^{n} y_r u_r(x) \geq u_{n+1}(x)$, $x \in S$,

DEFINITION 1. $u_1, u_2, \ldots, u_n$ are said to meet <u>Krein's conditions</u> if there are constants $c_1, c_2, \ldots, c_n$ such that

$$\sum_{r=1}^{n} c_r u_r(x) > 0, \quad x \in S.$$

We find immediately

LEMMA 2. If $u_1, u_2, \ldots u_n$ meet Krein's condition then $\tilde{D}$ is consistent. (It may well have an unbounded solution).

With $\tilde{D}$ we associate a <u>generalized moment</u>

problem which we denote as task $\tilde{P}$:

$$\text{Compute } \sup_{\alpha} \int_S u_{n+1}(x)d\alpha(x)$$

$$\text{subject to } \int_S u_r(x)d\alpha(x) = \mu_r, \; r = 1,2,\ldots,n$$

where $d\alpha$ is a positive measure.

DEFINITION 2. In R^j we define moment cones to be the subsets

$$M_j = \{v \mid v_r = \int_S u_r(x)d\alpha(x),\; r=1,2,\ldots,j;\; \begin{matrix} d\alpha \text{ positive and} \\ \text{bounded measure} \end{matrix}\},$$

$j = 1,2,\ldots,n+1$.

Thus P is consistent if and only if μ belongs to M_n.

LEMMA 3. M_j is the smallest cone which contains the set

$$C_j = \{u(x)=(u_1(x),u_2(x),\ldots,u_j(x)),\; x \in S\}.$$

PROOF. It is obvious that $C_j \subset M_j$. Assume now that $v \in M_j$. That is there is a positive measure $d\alpha$ such that

$$v_r = \int_S u_r(x)d\alpha(x), \quad r = 1,2,\ldots,j.$$

Put

$$a = \int_S d\alpha(x)$$

and $\bar{v}_r = v_r/a, \quad r = 1,2,\ldots,j$. We approximate the integrals

$$\bar{v}_r = \frac{1}{a}\int_S u_r(x)d\alpha(x) \qquad r = 1,2,\ldots,j$$

with Riemann-Stieltjes sums using the same partition for all j integrals. Since u_r, $r = 1,2,\ldots,j$ are continuous and S is compact we can approximate $\bar{v}_r$ arbitrarily close with sums of the form

$$\sum_j u(x^j)w_j, \quad \sum_j w_j = 1.$$

The latter belong to the convex hull of C_j and since C_j is compact its convex hull has the same property. Hence $\bar{v}$ must belong to the convex hull of C_j from which the conclusion follows.

LEMMA 4. If $u_1,u_2,\ldots,u_n$ meet Krein's condition then M_n is closed.

PROOF. An argument which is based on Helly's selection principle, can be carried through exactly as in [10, pp. 38-39].

We next prove

THEOREM 1. Let M_n and M_{n+1} be closed. Then we can assert: (if M_n is a proper subset of R_n)

a) $\tilde{D}$ has unbounded solutions if and only if $\tilde{P}$ is inconsistent.

b) If $\tilde{P}$ is consistent then $\tilde{P}$ and $\tilde{D}$ have equal optimal values. $\tilde{P}$ assumes its supremum value.

PROOF.

1. Let $d\alpha_0$ be a feasible solution of $\tilde{P}$, y^0 a feasible solution of $\tilde{D}$. Then

$$\int_S u_{n+1}(x)d\alpha_0(x) \leq \int_S \sum_{r=1}^{n} y_r^o u_r(x)d\alpha_0(x)$$

$$= \sum_{r=1}^{n} y_r^o \mu_r .$$

Hence if $\tilde{P}$ is feasible then $\tilde{D}$ cannot have unbounded solutions.

2. Now let $\tilde{P}$ be inconsistent. We want to show that $\tilde{D}$ has unbounded solutions. Thus μ is

outside of M_n. Hence there is a hyperplane H such that $H(z) \geq 0$, $z \in M_n$ and $H(\mu) = -d < 0$. Since M_n is a cone we can also require that H pass through the origin and write therefore

$$H(z) = \sum_{r=1}^{n} y_r z_r, \qquad z \in R^n.$$

In particular

$$\sum_{r=1}^{n} y_r u_r(x) \geq 0 \qquad x \in S$$

since the vectors $u(x)$, $x \in S$ belong to M_n. Now let $\tilde{y}$ be a feasible solution of D and put $\hat{y}(s) = \tilde{y} + sy$, $s > 0$. We verify directly that $\hat{y}(s)$ is a feasible solution. Further

$$\sum_{r=1}^{n} \mu_r \hat{y}_r(s) = \sum_{r=1}^{n} \tilde{y}_r \mu_r + s \sum_{r=1}^{n} y_r \mu_r = \sum_{r=1}^{n} \tilde{y}_r \mu_r - sd.$$

The value of this expression can be made arbitrarily small. Hence D has unbounded solutions. This concludes our proof of assertion a.

We next prove assertion b. We have earlier proven that $\tilde{P}$ has a bounded supremum. We denote it with λ^*. Consider the straight line $\{(\mu,\lambda)|-\infty<\lambda<\infty\}$. If $\lambda < \lambda^*$ then the point (μ,λ) is inside M_{n+1}. Since M_{n+1} is closed we can conclude that (μ,λ^*) belongs to M_{n+1}. We want to show that λ^* is also the optimal value of D.

Let $\varepsilon > 0$ be given and define β as the n+1-dimensional vector $(\mu,\lambda+\varepsilon)$.

Since β is outside of M_{n+1} there is a linear functional π such that $\pi(\beta) > 0$, $\pi(z) \leq 0$, $z \in M_{n+1}$.

Since $\pi(\beta) > \pi(\mu,\lambda^*)$ we can write

$$\pi(z) = z_{n+1} - \sum_{r=1}^{n} y_r z_r .$$

In particular

$$u_{n+1}(x) - \sum_{r=1}^{n} y_r u_r(x) \leq 0, \quad x \in S$$

$$\lambda^* + \varepsilon - \sum_{r=1}^{n} y_r \mu_r > 0$$

Hence

$$\sum_{r=1}^{n} y_r \mu_r < \lambda^*+\varepsilon \text{ and } \sum_{r=1}^{n} y_r u_r(x) \geq u_{n+1}(x) \qquad x \in S.$$

Since ε is arbitrary we see that λ^* is indeed the infimum-value of $\tilde{D}$. This concludes the proof of the theorem.

REMARK. We have shown that $\tilde{P}$ assumes its optimum if it is consistent. The corresponding statement is not true for $\tilde{D}$. Consider the example

$\tilde{P}$	$\tilde{D}$
$\sup_{\alpha} \int_0^1 \sqrt{t}\, d\alpha(t)$	$\inf\ y_1$
$\int_0^1 d\alpha(t) = 1$	$y_1 + y_2 t \geq \sqrt{t} \quad t \in [0,1]$
$\int_0^1 t\,d\alpha(t) = 0$	
$\alpha \nearrow$	

The common objective value is 0 and $\tilde{P}$ has as its only feasible solution the mass 1 situated at the origin. But if we take $y_1 = \varepsilon > 0$ we must put $y_2 = 0.25/\varepsilon$. Hence $(\tilde{D})$ cannot assume its inf. value 0.

THEOREM 2. $\tilde{P}$ assumes its maximal value for a point mass distribution with at most n masspoints.

PROOF. From the argument used to prove the assertion b of Theorem 1, it follows that the maximal value, λ^* of $\tilde{P}$, can be associated with the vector (μ,λ^*) in M_{n+1}. According to Lemma 3 this cone is the conic hull of the curve

$$(u_1(x), u_2(x), \ldots, u_{n+1}(x)), \qquad x \in S.$$

From the Carathéodory theorem we can conclude that (μ,λ^*) admits a representation

$$\lambda^* = \sum_{j=1}^{n+2} m_j u_{n+1}(x^j), \tag{3}$$

$$\mu_r = \sum_{j=1}^{n+2} m_j u_r(x^j), \qquad r = 1,2,\ldots,n, \tag{4}$$

$$m_j \geq 0 \qquad j = 1,2,\ldots,n+2.$$

Equations (3) and (4) can, for $x^1, x^2, \ldots, x^{n+2}$ fixed be considered as a linear program with the variables as $m_1, m_2, \ldots, m_{n+2}$. When we maximize over $m_1, m_2, \ldots, m_{n+2}$ we cannot get a larger optimal value than λ^*. But the program has an optimal solution with at most n positive masses. This gives the desired representation.

We are now ready to extend our results to moment problems with constraints on y and prove

THEOREM 3. Consider the tasks

$\tilde{D}_F$: Compute $\inf \sum_{r=1}^{n} y_r \mu_r$

$$\text{subject to } \sum_{r=1}^{n} y_r u_r(x) \geq u_{n+1}(x), \qquad x \in S$$

$$|y_r| \leq F, \qquad r = 1,2,\ldots,n$$

$\tilde{P}_F$: Compute $\max_{\alpha} \int_0^1 u_{n+1}(x)\,d\alpha(x) - F \sum_{r=1}^{n} (m_r^+ + m_r^-)$

$$\int_S u_r(x)\,d\alpha(x) + m_r^+ - m_r^- = \mu_r$$

$$r = 1,2,\ldots,n$$

$$m_r^+ \geq 0,\ m_r^- \geq 0,\ r = 1,2,\ldots,n$$

$d\alpha$ a positive measure.

Then the following statements hold true:

1. $\tilde{P}_F$ is bounded if and only if $\tilde{D}_F$ is feasible.

2. If bounded, $\tilde{P}_F$ and $\tilde{D}_F$ have the same optimal value λ^*.

3. If $u_1, u_2, \ldots, u_n$ meet Krein's condition then λ^* admits a representation

$$\lambda^* = \sum_{i=1}^{n} m_i u_{n+1}(x^i) - F \sum_{r=1}^{n} (m_r^+ + m_r^-)$$

$$\mu_r = \sum_{i=1}^{n} m_i u_r(x^i) + m_r^+ - m_r^- \qquad r = 1,2,\ldots,n$$

$$m_i \geq 0 \quad i = 1,2,\ldots,n, \; m_r^+ \geq 0, \; m_r^- \geq 0$$

where at most n of the numbers m_i, $i = 1,2,\ldots,n$ m_r^+, m_r^-, $r = 1,2,\ldots,n$ are positive and $m_r^+ \cdot m_r^- = 0$, $r = 1,2,\ldots,n$.

PROOF. The argument is analogous to that used in verifying Theorems 1 and 2. Instead of M_j we use the smallest cones containing the vectors

$$u(x) = (u_1(x), u_2(x), \ldots, u_j(x)), \qquad x \in S$$

and

$$\pm e^r \qquad r = 1,2,\ldots,n$$

where the k-th entry e_k^r of e^r is given by

$e_k^r = \delta_{kr} + F \cdot \delta_{kn+1}$ (δ is Kronecker's delta).

Now we can establish

LEMMA 5 (Complementary slackness). Let $d\alpha_0$ be an optimal solution of $\tilde{P}$ and y^0 an optimal solution of $\tilde{D}$. Then

$$\int_S \left(\sum_{r=1}^{n} y_r^0 u_r(x) - u_{n+1}(x)\right) \, d\alpha_0(x) = 0.$$

PROOF. The optimal values of $\tilde{P}$ and $\tilde{D}$ are equal. Hence

$$\int_S u_{n+1}(x) \, d\alpha_0(x) = \sum_{r=1}^{n} y_r \mu_r.$$

But

$$\int_S u_r(x) \, d\alpha_0(x) = \mu_r \qquad r = 1,2,\ldots,n$$

and thus

$$\int_S \left(\sum_{r=1}^{n} y_r^0 u_r(x) - u_{n+1}(x)\right) d\alpha_0(x) = 0.$$

From the fact that $\sum_{r=1}^{n} y_r^0 u_r(x) \geq u_{n+1}(x)$ and $d\alpha_0$ is a positive measure, the assertion follows.

Thus $\sum_{r=1}^{n} y_r^0 u_r$ interpolates u_{n+1} at the mass-points of $d\alpha_0$. We are now ready to prove

THEOREM 4. With problem D:

$$\text{Compute } \min_{y} G(y)$$

subject to

$$\sum_{r=1}^{n} y_r u_r(x) \geq u_{n+1}(x), \qquad x \in S,$$

we can associate the nonlinear system

$$\text{(5)} \qquad \sum_{i=1}^{q} m_i u_r(x^i) + m_r^+ - m_r^- = \frac{\partial F}{\partial y_r}, \quad r = 1,2,\ldots,n$$

$$\text{(6)} \qquad \sum_{r=1}^{n} y_r u_r(x^i) = u_{n+1}(x^i), \quad i = 1,2,\ldots,q$$

with the side-conditions

$$m_r^+(F+y_r) = 0, \qquad r = 1,2,\ldots,n$$

$$m_r^-(F-y_r) = 0, \qquad r = 1,2,\ldots,n$$

$$m_i > 0 \qquad i = 1,2,\ldots,q$$

$$m_r^+ \geq 0,\ m_r^- \geq 0, \qquad r = 1,2,\ldots,n$$

where y is an optimal solution of D.

PROOF. Let y* be an optimal solution of D. Let D* denote the linear semi-infinite program which is obtained by replacing the objective function D by the linear expression

$$\sum_{r=1}^{n} y_r \left(\frac{\partial G}{\partial y_r}\right)_{y=y^*}.$$

Then y is an optimal solution of D if and only if it is an optimal solution of D* (Lemma 1). Hence we can consider D* together with the corresponding moment problem and apply Lemma 4 to this dual pair of linear semi-infinite programs. If we now put y* = y in D* we obtain the desired equations.

THEOREM 5. If $u_1, u_2, \ldots, u_{n+1}$ have continuous partial derivatives of the first order, x^j is in the interior of S, y is a feasible solution of D_F and

$$\sum_{r=1}^{n} y_r u_r(x^j) = u_{n+1}(x^j),$$

then

$$\nabla\left(\sum_{r=1}^{n} y_r u_r - u_{n+1}\right)_{x=x^j} = 0. \tag{7}$$

PROOF. Since y is feasible the function

$$\Psi = \sum_{r=1}^{n} y_r u_r - u_{n+1}$$

is nonnegative in S. Further, x^j is in the interior which implies that there is a vector h such that $x^j - h$, $x^j + h$ both lie in S. Put $g(\lambda) = \Psi(x^j+\lambda h)$, $-1 \leq \lambda \leq 1$. Thus g is nonnegative over [-1,1] and $g(0) = 0$. We must conclude $g'(0) = 0$.

Since x^j is the interior of S the construction indicated above can be done for linearly independent vectors $h^1, h^2, \ldots, h^k$. (Recall that S is a subset of R^k). From this the assertions follows.

Thus (7) can be combined with (5) and (6) to form a nonlinear system (system NL of [8]) from whose solutions, an objective vector of D, can be found.

REMARK. If x^j is a boundary point then some, but not k, independent vectors $h^1, h^2 \ldots$ may be found such that $x^j - h^s$, $x^j + h^s$ belong to S. This gives conditions which should be included in the nonlinear system.

4. Numerical solution.

The general strategy is to construct an approximate solution to the nonlinear system (5), (6) and (7) which is then solved numerically.

We shall now show how an arbitrarily good approximation can be constructed by means of linear programming. Strict upper and lower bounds can also be obtained in the process.

The main idea is to approximate the moment generating functions $u_1, u_2, \ldots, u_{n+1}$, as well as the preference function G, piecewise with linear functions.

Let $T = \{x^1, x^2, \ldots, x^N\}$ and $\Gamma = \{\eta^1, \eta^2, \ldots, \eta^\ell\}$ be finite subsets of S and K respectively. We shall construct approximations to $u_1, u_2, \ldots, u_{n+1}$ and G using functional values on T and Γ. Following [8] we shall call T and Γ grids. We prove

LEMMA 6. Let $\delta_1, \delta_2, \ldots, \delta_N$ be non-negative functions on S. For any function Ψ we define the non-negative linear operator L_T by

$$L_T \Psi(x) = \sum_{j=1}^{N} \delta_j(x) \Psi(x^j).$$

Then y is a solution of the set of inequalities

$$\sum_{r=1}^{n} y_r L_T u_r(x) \geq L_T u_{n+1}(x), \qquad x \in S \tag{8}$$

if and only if y is a solution of

$$\sum_{r=1}^{n} y_r u_r(x^j) \geq u_{n+1}(x^j), \qquad x^j \in T. \tag{9}$$

PROOF. For fixed x the inequality (8) is a positive linear combination of (9). Hence the desired result is a consequence of Farkas' Lemma.

The following examples illustrate the use of Lemma 6.

1. Let x be a convex combination of a subset of k+1 elements in $T \subset S \subseteq R^k$, say

$$x = \sum_{j=1}^{k+1} \lambda_j x^j, \quad \sum_{j=1}^{k+1} \lambda_j = 1, \quad \lambda_j \geq 0.$$

Put

$$(L_T \Psi)(x) = \sum_{j=1}^{k+1} \lambda_j \psi(x^j).$$

We note that $L_T\Psi = \Psi$ if Ψ is a linear function.

2. Define k strictly increasing sequences $c_1, c_2, \ldots, c_k$ where $c_j = (x_{j1}, x_{j2}, \ldots, x_{jv_j})$, $j = 1,2,\ldots,k$. Now let T be a set with $v_1 \cdot v_2 \cdot \ldots \cdot v_k$ points represented by vectors of the form $(x_{j_1 1}, x_{j_2 2}, \ldots, x_{j_k k})$ where $x_{j_s s}$ is an element of the sequence c_s. We assume also that S is contained in the k-dimensional rectangle defined by the Cartesian product

$$[x_{11}, x_{1v_1}]\times[x_{21}, x_{2v_2}]\times\ldots\times[x_{k1}, x_{kv_k}].$$

If x belongs to S we can find a subrectangle

$$[x_{1j_1}, x_{1j_1+1}]\times[x_{2j_2}, x_{2j_2+1}]\times\ldots\times[x_{kj_k}, x_{kj_k+1}]$$

containing x. We then define $\Psi_T(x)$ as the result of repeated one-dimensional linear interpolations using the values of Ψ in the corners of the subrectangle.

We note that for $k = 2$, $\Psi_T(x) = \Psi(x)$ if Ψ is of the form $\Psi(x) = a_1x_1x_2 + a_2x_1 + a_3x_2 + a_4$

where a_j are constants.

We next discuss approximation of G. We can define L_Γ for functions on K analogously with L_T for functions on S. Since G is convex we can easily construct one-sided piecewise linear approximation for G(y).

Let $\Gamma = \{\eta^1, \eta^2, \ldots, \eta^\ell\}$ be a grid in K. For each j we construct π_j the supporting plane of G at η^j. Then

$$G(y) \geq \max_j \pi_j(y), \qquad y \in K. \tag{10}$$

If y also is a convex combination of elements in Γ and we define $L_\Gamma \Psi$ as in example 1 after Lemma 6 we find

$$(L_\Gamma G)(y) \geq G(y).$$

THEOREM 6. Let z be the optimal value of problem D, T and Γ grids in S and K respectively. Then

$$z \geq z_0 = \min y_0$$

subject to $y_0 - \pi_v(y) \geq 0 \qquad \eta^v \in \Gamma$

(11) $\quad \sum_{r=1}^{n} y_r u_r(x^j) \geq u_{n+1}(x^j) \qquad x^j \in T$

(12) $\quad |y_r| \leq F, \qquad r = 1,2,\ldots,n.$

PROOF. The result follows immediately from the fact that $G(y) \geq \pi_v(y)$, $\eta^v \in \Gamma$ and that if y is a feasible solution of D then it certainly satisfies (11) and (12).

REMARK 1. y_0 is the solution of a linear program with n+1 variables subject to $2n + \ell + N$ constraints where ℓ is the number of elements in Γ, N the number of elements in T.

REMARK 2. Using Lemma 6 we realize that if y is an optimal solution of the linear program of Theorem 6 then y is also an optimal solution of D if $L_\Gamma G = G$ and $L_T u_r = u_r$, $r = 1,2,\ldots,n+1$. If we refine the grids we can make $L_\Gamma G - G$ and $L_T u_r - u_r$ small if G and u_r are continuous functions and L_T and L_Γ are constructed as indicated.

REMARK 3. An upper bound for the optimal

value of D is given by G(y) for any feasible y. A systematic choice can be made as follows. Let

$$\kappa = \sum_{r=1}^{n} c_r u_r$$

be a strictly positive function and put

$$\gamma_0 = \inf \kappa(x), \qquad x \in S.$$

Further, let $y^1 = (y_0^1, y_1^1, \ldots, y_n^1)$ be an optimal solution of the linear program in Theorem 6. Define

$$\Delta = \min \sum_{r=1}^{n} y_r^1 u_r(x) - u_{n+1}(x), \; x \in S.$$

Then y^2 is a feasible solution of D where

$$y_r^2 = y_r^1 + \frac{\Delta}{\gamma_0} c_r, \qquad r = 1,2,\ldots,n,$$

and we find that the optimal value of D is contained in the interval $[G(y^1), G(y^2)]$. We note that Δ can be made small if the grid is fine. See [8] also.

We now establish a general convergence result:

THEOREM 7. Consider the semi-infinite program D (see the beginning of §1) and assume that the set of feasible solutions have interior points. Denote by K_F the set $\{y \mid |y_r| \leq F\}$. We consider now a sequence of continuous function systems $u_1^j, u_2^j, \ldots, u_n^j$, $j = 1,2,\ldots$ converging uniformly on S towards $u_1, u_2, \ldots, u_n$ and a sequence of convex continuous functions $G_1, G_2, \ldots$ converging to G, uniformly on K_F. We form the sequence of programs $D_1, D_2, \ldots$

$$\min_y \quad G_j(y)$$

$$\text{subject to} \quad \sum_{r=1}^{n} y_r u_r^j(x) \geq u_{n+1}^j(x), \qquad x \in S$$

$$|y_r| \leq F, \qquad r = 1,2,\ldots,n.$$

Then

1. There is an index j_0 such that D_j is feasible for $j \geq j_0$.
2. If y^j is an optimal solution of D_j, then the accumulation points of $\{y^j\}_{j_0}^{\infty}$ are optimal solutions of D.

PROOF. Let y be a vector, which is interior

in the set of feasible solutions of D. Then

$$\gamma = \min_{x \in S} \sum_{r=1}^{n} y_r^0 u_r(x) - u_{n+1}(x)$$

is strictly positive. Due to the uniform convergence we can conclude that there is a number j_0 such that y^0 is a feasible solution of D_j, $j > j_0$.

Now let y^* be an accumulation point of $\{y^j\}_{j_0}^{\infty}$. Then there is a subsequence converging towards y^*. We denote its elements by $\{y^{\ell(j)}\}$. $y^{\ell(j)}$ is a feasible solution of $D_{\ell(j)}$. Hence

$$\sum_{r=1}^{n} y_r^{\ell(j)} u_r^{\ell(j)}(x) \geq u_{n+1}^{\ell(j)}(x), \quad x \in S.$$

Therefore

$$\sum_{r=1}^{n} y_r^* u_r(x) \geq u_{n+1}(x), \qquad x \in S,$$

and y^* is thus a feasible solution of D. Denote its optimal value by z. We find

$$z \leq G(y^*).$$

Let $\tilde{y}$ be an optimal solution vector of D. $\tilde{y}$ is a feasible solution of D. Therefore we can construct a set $\{\eta^j\}$ of vectors such that η^j is a feasible solution of D_j and $\eta^j \to \tilde{y}$ when $j \to \infty$. Hence $G_{\ell(j)}(\eta^{\ell(j)}) \geq G_{\ell(j)}(y^{\ell(j)})$ giving $G(\tilde{y}) \geq G(y^*)$ or

$$z \geq G(y^*).$$

Thus $z = G(y^*)$. The proof is concluded.

Theorem 7 is applicable for example, when we approximate u_r by means of higher order interpolation. This may be an alternative to refinement of the grid T if we want to improve the accuracy of an approximate solution. We note however, that the corresponding semi-infinite programs cannot be solved directly by means of linear programming. Furthermore, the functions used to approximate G must be convex, in order to assure that the techniques discussed here apply.

The solution of an approximating program can be used to construct an initial approximation for the solution of the system (5), (6) and (7). A difficulty has to be resolved. We estimate q (and hence the structure of the system to be solved) by means of the zeros of the functions

$$\Psi_j = \sum_{r=1}^{n} y_r^j u_r^j - u_{n+1}^j, \qquad j = 1,2,\ldots$$

where y^j is an optimal solution of an approximating program D_j. But the number of zeros may change with j. Consider the particular example

$$\Psi_j(t) = (t^2+\frac{1}{j})(1-\frac{1}{j}+t)(1-\frac{10}{\ln j}-t)^2,$$

$$j = 2,3,\ldots$$

$$\Psi(t) = t^2(1+t)(1-t)^2$$

defined on [-1,1].

We notice that the limit-function has a double root for t = 0 which does not appear in Ψ_j for any j.

Now let $\Psi_1, \Psi_2, \ldots$ be a sequence of functions, non-negative and continuous on S and converging uniformly towards Ψ. With Ψ_j we associate

$Z_j = \{x \mid x \in S, \Psi_j(x) = 0\}$. Assume Z_j is not empty, $j = 1,2,\ldots$ and take $z^j \in Z_j$. Let z^* be an accumulation point of $\{z^j\}_1^\infty$. (At least one such point must exist since S is compact). Then we easily verify that $\Psi(z^*) = 0$.

Consider now the sequence of programs defined in the proof of Theorem 7. Let y^j be the optimal solution of program D_j. We can, without loss of generality, assume that $y^j \to y$. Put

$$\Psi_j = \sum_{r=1}^{n} y_r^j u_r^j - u_{n+1}^j$$

$$\Psi = \sum_{r=1}^{n} y_r u_r - u_{n+1}$$

Define Z_j as above and denote by Z the set of zeros of Ψ. We observe that $\{\Psi_j\}$ converges towards ψ uniformly on S. For program D we form the nonlinear system obtained by combining (5), (6) and (7), assuming that necessary derivatives do exist. Then the numbers $x^1,x^2,\ldots,x^q$ belong to Z. We can write Z as $Z_0 \cup Z_1$ where Z_0 is the set of accumulation points of all sequences of the form $\{z^j\}$, $z^j \in Z_j$.

From the fact that

$$G(y) = \sum_i m_i u_{n+1}(x^i), \quad x^i \in Z,$$

and

$$G^j(y^j) = \sum_i m_i^j u_{n+1}^j(x^{i(j)}), \; x^{i(j)} \in Z_j,$$

follows that we need only consider x^i in Z_0.

We can conclude that the following numerical procedure is valid.

1. Form a sequence of approximating convex programs as described in the proof of Theorem 7. (If we use the procedure given in Theorem 6, these programs can be handled by means of linear programming).
2. Take q as the number of accumulation points of the sequences x^j formed as discussed above. (Several distinct sequences may have the same accumulation points. A sufficient condition for this not being the case is given in [8]).

3. Construct the system (5), (6) and (7) in accordance with the adopted approximations for $x^1, x^2, \ldots, x^q$.

4. Solve this system by numerical methods.

REFERENCES

[1] A. Charnes and W. W. Cooper, Management Models and Industrial Applications of Linear Programming, Vols. I and II, J. Wiley and Sons, N.Y., 1961.

[2] A. Charnes, W. W. Cooper and K. O. Kortanek, Duality, Haar programs and finite sequence spaces, Proc. Nat. Acad. Sci. U.S.A., 48 (1962) 783-786.

[3] A. Charnes, W. W. Cooper and K. O. Kortanek, On the theory of semi-infinite programming and a generalization of the Kuhn-Tucker saddle point theorem for arbitrary convex functions, Nav. Res. Logist. Quart. 16 (1969) 41-51.

[4] G. B. Dantzig, Linear Programming and Extensions, Princton University Press, 1963.

[5] H. G. Fortak, Numerical simulation of the temporal and spatial distributions of urban air pollution concentration, Proc. Symp. Multiple-Source Urban Diffusion Models, Research Triangle Park, North Carolina, Nov. 1970.

[6] S. -Å. Gustafson, On the computational solution of a class of generalized moment problems, SIAM J. Numer. Anal., 7 (1970) 343-357.

[7] S. -Å. Gustafson and K. O. Kortanek, Analytical properties of some multiple-source urban diffusion models, Environment and Planning, 4 (1972) 31-41.

[8] S. -Å. Gustafson and K. O. Kortanek, Numerical treatment of a class of semi-infinite programming problems, IPP Report No. 21, School of Urban and Public Affairs, Carnegie-Mellon Univ., Pittsburgh, Aug. 1971.

[9] S. -Å. Gustafson, K. O. Kortanek and W. Rom, Non-Cebysevian moment problems, SIAM J. Numer. Anal., 7 (1970) 335-342.

[10] S. Karlin and W. J. Studden, Tchebycheff systems with applications in analysis and statistics, Interscience Publishers, N.Y., 1966.

[11] K. O. Kortanek and J. P. Evans, Pseudo-concave programming and Lagrange regularity, Operations Res., 15 (1967) 882-891.

[12] D. M. Rote, J. W. Gudenas and L. A. Cooley, Studies of the Argonne integrated-puff model, Argonne National Laboratory, Ill., October 1971.

[13] D. H. Slade (Editor), Meteorology and atomic energy 1968, TlD-24190, U.S. Atomic Energy Commission, July 1968.

ON THE SOLUTION OF LARGE SYSTEMS OF LINEAR ALGEBRAIC EQUATIONS WITH SPARSE, POSITIVE DEFINITE MATRICES

David M. Young

1. Introduction.

The object of this paper is to discuss the current status of iterative methods for solving large systems of linear algebraic equations. Primary emphasis is on those systems involving sparse matrices where iterative methods appear more attractive than direct methods. However, we note that the size of problems for which direct methods should be used has increased substantially over the past few years with advances in computing machinery.

An example of a problem leading to a large

Work on this paper was supported in part by Grant DA-ARO-D-31-124-72-G34 (Army Research Office, Durham) and NSF Grant GP-23655 at the University of Texas at Austin.

linear system with a sparse matrix is the following. Given a continuous function $g(x,y)$ defined on the boundary of the unit square $0 \leq x \leq 1$, $0 \leq y \leq 1$, we seek to determine a twice differentiable function $u(x,y)$ defined in the closed square, satisfying Laplace's equation in the interior of the square, and coinciding with $g(x,y)$ on the boundary. We shall refer to this problem as the _model problem_ [29].

We consider the following procedure for solving the model problem. Given a positive integer J, we cover the square with a network of uniform mesh size $h = 1/J$. We seek to find approximate values of $u(x,y)$ at the interior nodes. (Those on the boundary are already known, of course.) We replace the Laplace differential equation

$$(1.1) \qquad \frac{\partial^2 u}{\partial x^2} + \frac{\partial^2 u}{\partial y^2} = 0$$

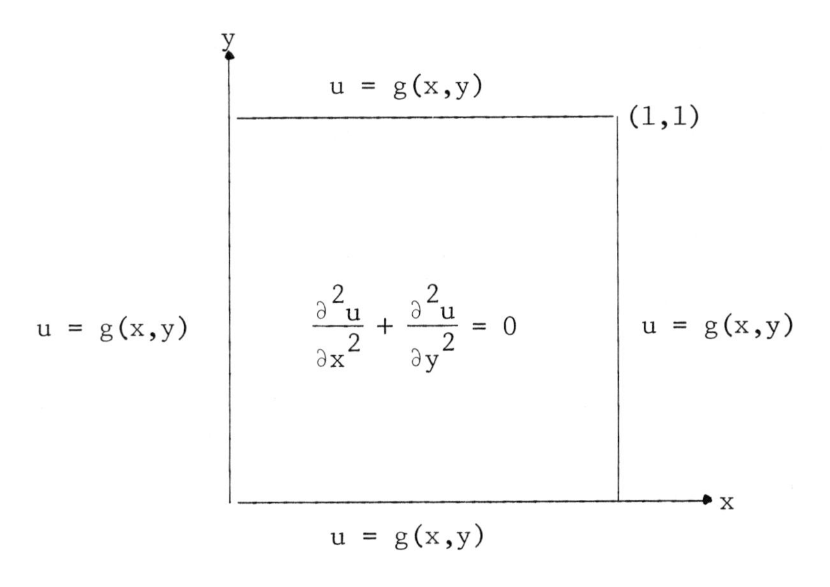

Figure 1.1. The Model Problem

by the usual 5-point difference equation

$$(1.2) \qquad 4u(x,y)-u(x+h,y)-u(x,y+h)-u(x-h,y)-u(x,y-h)=0.$$

This equation is obtained by representing $\frac{\partial^2 u}{\partial x^2}$ by

$$(1.3) \qquad \frac{u(x+h,y) + u(x-h,y) - 2u(x,y)}{h^2}$$

and $\frac{\partial^2 u}{\partial y^2}$ by a similar expression, and then substituting in the differential equation and multiplying by $-h^2$.

In the case $h = 1/3$ we have the four equations (see Figure 1.2)

$$(1.4)\quad \begin{cases} 4u_1 - u_2 - u_3 = u_6 + u_{16} \\ -u_1 + 4u_2 - u_4 = u_7 + u_9 \\ -u_1 + 4u_3 - u_4 = u_{13} + u_{15} \\ -u_2 - u_3 + 4u_4 = u_{10} + u_{12} \end{cases}$$

where u_i is the value of the unknown function $u(x,y)$ at the i-th node with the nodes labeled as in Figure 1.2.

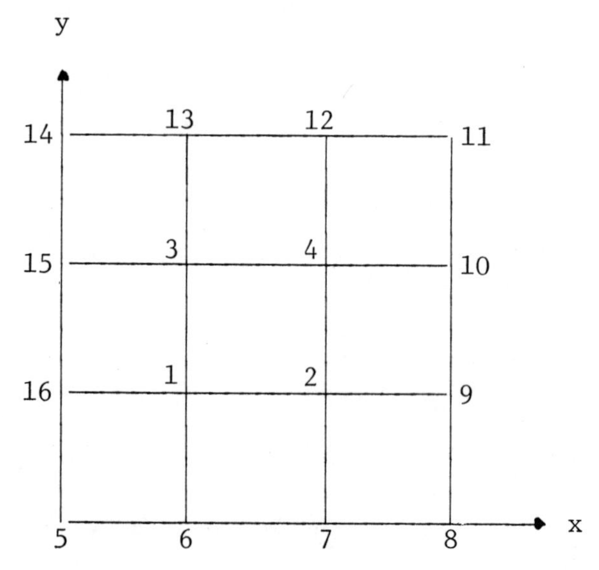

Figure 1.2. Numerical Solution of the Model Problem

It is easy to show that the coefficient matrix A of the system (1.4) is a positive definite matrix and hence the system has a unique solution.

If h^{-1} is very large, say of the order of 100 - 500, the system is very large; however, the matrix is very sparse since there are never more than 5 non-zero elements in any row. Thus, the use of iterative methods seems indicated.

In Section 2 we discuss various iterative methods including the Jacobi method, the Gauss-Seidel method, the successive overrelaxation method, and the symmetric successive overrelaxation method. In Section 3 we study the convergence of these methods under various assumptions on A. A "standard of convergence" is established and used as a basis for evaluating the various methods. In Section 4 we consider the use of semi-iterative methods and second-degree methods to accelerate the convergence of some methods. In particular, it is shown that one can obtain an order-of-magnitude improvement over this standard of convergence even if <u>no assumption</u> is made on A beyond positive definiteness.

In Section 5 we discuss the application of

these methods to solve boundary value problems involving elliptic partial differential equations. Other methods, which give substantially faster convergence, but in very restricted classes of cases, than the "basic" methods are discussed in Section 6. These include alternating direction implicit methods and quasi-direct methods.

Methods for solving linear systems can be applied to nonlinear systems in several ways. First, as shown in Section 2, several of the methods can be applied directly to nonlinear systems. In order to carry out a generalized method it is necessary to solve a sequence of problems each involving the determination of the solution of a single equation with one unknown. It is hoped that, at least for nearly linear systems, some of the properties of the methods which can be proved to hold for linear systems will apply, at least reasonably well, for nonlinear systems.

Another way in which methods designed for linear systems can be used for nonlinear systems is in the use of an iterative procedure, each step of which involves the solution of a system of linear algebraic equations [19]. As an example, consider the system

$$\text{(1.5)}\qquad \begin{cases} 4u_1 - u_2 + \frac{1}{10}e^{u_1} = 1 \\ -u_1 + 4u_2 + \frac{1}{8}u_1^2 = 0. \end{cases}$$

One could consider the iterative method defined by

$$\text{(1.6)}\qquad \begin{cases} 4u_1^{(n+1)} - u_2^{(n+1)} + \frac{1}{10}e^{u_1^{(n)}} = 1 \\ -u_1^{(n+1)} + 4u_2^{(n+1)} + \frac{1}{8}(u_1^{(n)})^2 = 0. \end{cases}$$

To determine $u_1^{(n+1)}$ and $u_2^{(n+1)}$ from $u_1^{(n)}$ and $u_2^{(n)}$ involves solving a system for two linear equations with two unknowns. This could be done directly, of course. However, if one chose to use an iterative method the iterations involved would be considered as "inner iterations." The process of going from $u^{(n)}$ to $u^{(n+1)}$ would be an "outer" iteration. Clearly, the better the inner iteration process, the better the overall method.

Another (less trivial) example is the Newton method for solving a nonlinear system. As a simple case consider the system

$$(1.7)\qquad \begin{cases} f_1(u_1,u_2) = 0 \\ f_2(u_1,u_2) = 0 \end{cases}$$

which we may write as

$$(1.8)\qquad Fu = 0.$$

The Newton method is defined by

$$(1.9)\qquad u^{(n+1)} = u^{(n)} - [F'u^{(n)}]^{-1}Fu^{(n)}$$

where $F'u^{(n)}$ is the Jacobian

$$(1.10)\qquad F'u^{(n)} = \begin{pmatrix} \dfrac{\partial f_1}{\partial u_1} & \dfrac{\partial f_1}{\partial u_2} \\ \\ \dfrac{\partial f_2}{\partial u_1} & \dfrac{\partial f_2}{\partial u_2} \end{pmatrix}$$

where the partial derivatives are evaluated at $(u_1^{(n)},u_2^{(n)})$. (See, for instance [20].) Actually, rather than compute $[F'u]^{-1}$ we obtain $u^{(n+1)}$ by solving the linear system

$$(1.11) \qquad [F'u^{(n)}]u^{(n+1)} = [F'u^{(n)}]u^{(n)} - Fu^{(n)}.$$

Another example would involve splitting the nonlinear operator F appearing in (1.8) in the form

$$(1.12) \qquad F = F_1 + F_2$$

where F_1 is linear. The iterative method can be given as

$$(1.13) \qquad F_1u^{(n+1)} + F_2u^{(n)} = 0.$$

Thus in the example (1.5) we have

$$(1.14) \qquad F_1u = \begin{pmatrix} 4u_1 - u_2 \\ -u_1 + 4u_2 \end{pmatrix}$$

$$(1.15) \qquad F_2u = \begin{pmatrix} \frac{1}{10} e^{u_1} - 1 \\ \frac{1}{8} u_1^2 \end{pmatrix}$$

Another example would arise in the numerical solution of the differential equation

$$(1.16) \qquad \frac{\partial^2 u}{\partial x^2} + \frac{\partial^2 u}{\partial y^2} + \frac{1}{10} e^u = 0.$$

Here one could let F_1 correspond to the discrete representation of

$$(1.17) \qquad \frac{\partial^2 u}{\partial x^2} + \frac{\partial^2 u}{\partial y^2}$$

and F_2 could correspond to

$$(1.18) \qquad \frac{1}{10} e^u .$$

2. The Basic Iterative Methods.

In this section we define four "basic" iterative methods, namely the Jacobi, Gauss-Seidel, successive overrelaxation, and symmetric successive overrelaxation methods. We first define the methods as applied to the linear system

$$(2.1) \qquad Au = b$$

where A is a real $N \times N$ matrix and b is a real $N \times 1$

column vector. While in our later discussion we will be assuming that A is positive definite,* for this section we only assume that the diagonal elements of A do not vanish. We also define each method for the nonlinear system

$$Fu = 0 \tag{2.2}$$

where

$$Fu = \begin{pmatrix} f_1(u_1,u_2,\ldots,u_N) \\ f_2(u_1,u_2,\ldots,u_N) \\ \cdot\ \cdot\ \cdot\ \cdot\ \cdot\ \cdot\ \cdot\ \cdot\ \cdot \\ f_N(u_1,u_2,\ldots,u_N) \end{pmatrix} \tag{2.3}$$

We illustrate the methods for the case of three equations and three unknowns. For the linear system

---

*By stating that a real matrix is positive definite we mean that it is symmetric and has positive eigenvalues.

$$(2.4)\quad \begin{cases} a_{11}u_1 + a_{12}u_2 + a_{13}u_3 = b_1 \\ a_{21}u_1 + a_{22}u_2 + a_{23}u_3 = b_2 \\ a_{31}u_1 + a_{32}u_2 + a_{33}u_3 = b_3 \end{cases}$$

we first define the Jacobi method by

$$\begin{cases} a_{11}u_1^{(n+1)} + a_{12}u_2^{(n)} + a_{13}u_3^{(n)} = b_1 \\ a_{21}u_1^{(n)} + a_{22}u_2^{(n+1)} + a_{23}u_3^{(n)} = b_2 \\ a_{31}u_1^{(n)} + a_{32}u_2^{(n)} + a_{33}u_3^{(n+1)} = b_3. \end{cases}$$

Note that to get $u_i^{(n+1)}$ one in effect solves the i-th equation for u_i in terms of the available values of $u_j^{(n)}$ for $j \neq i$. Note also that values of $u_j^{(n)}$ for $j < i$ are used even though at the stage of improved values $u_j^{(n+1)}$ are available. For the nonlinear system

$$(2.5)\quad \begin{cases} f_1(u_1,u_2,u_3) = 0 \\ f_2(u_1,u_2,u_3) = 0 \\ f_3(u_1,u_2,u_3) = 0 \end{cases}$$

we find $u_1^{(n+1)}$, $u_2^{(n+1)}$, $u_3^{(n+1)}$ by

$$(2.6)\quad \begin{cases} f_1(u_1^{(n+1)}, u_2^{(n)}, u_3^{(n)}) = 0 \\ f_2(u_1^{(n)}, u_2^{(n+1)}, u_3^{(n)}) = 0 \\ f_3(u_1^{(n)}, u_2^{(n)}, u_3^{(n+1)}) = 0. \end{cases}$$

At each step one has to solve a single nonlinear equation for one unknown.

The Gauss-Seidel method is the same as the Jacobi method except that at each stage one uses the latest available value. Thus we have in the linear case

$$(2.7)\quad \begin{cases} a_{11}u_1^{(n+1)} + a_{12}u_2^{(n)} + a_{13}u_3^{(n)} = b_1 \\ a_{21}u_1^{(n+1)} + a_{22}u_2^{(n+1)} + a_{23}u_3^{(n)} = b_2 \\ a_{31}u_1^{(n+1)} + a_{32}u_2^{(n+1)} + a_{33}u_3^{(n+1)} = b_3 \end{cases}$$

and in the nonlinear case

$$(2.8)\quad \begin{cases} f_1(u_1^{(n+1)}, u_2^{(n)}, u_3^{(n)}) = 0 \\ f_2(u_1^{(n+1)}, u_2^{(n+1)}, u_3^{(n)}) = 0 \\ f_3(u_1^{(n+1)}, u_2^{(n+1)}, u_3^{(n+1)}) = 0. \end{cases}$$

The successive overrelaxation method (SOR method) is a slight modification of the Gauss-Seidel method. One first chooses a relaxation factor ω, usually between 0 and 2. Given $u^{(n)}$ one computes the tentative value $\tilde{u}_1^{(n+1)}$ by the Gauss-Seidel method and then accepts $u_1^{(n+1)}$ given by

$$(2.9) \qquad u_1^{(n+1)} = u_1^{(n)} + \omega(\tilde{u}_1^{(n+1)} - u_1^{(n)}).$$

Next, one computes $\tilde{u}_2^{(n+1)}$ using the Gauss-Seidel method, with $u_1^{(n+1)}, u_3^{(n)}, \ldots, u_N^{(n)}$ and then accepts $u_2^{(n+1)}$, which is obtained from

$$(2.10) \qquad u_2^{(n+1)} = u_2^{(n)} + \omega(\tilde{u}_2^{(n+1)} - u_2^{(n)}).$$

The procedure is continued until $u_1^{(n+1)}, u_2^{(n+1)}, \ldots, u_N^{(n+1)}$ are found. To obtain a set of formulas not involving $\tilde{u}_1^{(n+1)}$, $\tilde{u}_2^{(n+1)}$, $\tilde{u}_3^{(n+1)}$, we eliminate these quantities from the six equations

$$(2.11a) \qquad \begin{cases} a_{11}\tilde{u}_1^{(n+1)} + a_{12}u_2^{(n)} + a_{13}u_3^{(n)} = b_1 \\ u_1^{(n+1)} = u_1^{(n)} + \omega(\tilde{u}_1^{(n+1)} - u_1^{(n)}) \end{cases}$$

$$(2.11b)\quad \begin{cases} a_{21}u_1^{(n+1)} + a_{22}\tilde{u}_2^{(n+1)} + a_{23}u_3^{(n)} = b_2 \\ u_2^{(n+1)} = u_2^{(n)} + \omega(\tilde{u}_2^{(n+1)} - u_2^{(n)}) \end{cases}$$

$$(2.11c)\quad \begin{cases} a_{31}u_1^{(n+1)} + a_{32}u_2^{(n+1)} + a_{33}\tilde{u}_3^{(n+1)} = b_3 \\ u_3^{(n+1)} = u_3^{(n)} + \omega(\tilde{u}_3^{(n+1)} - u_3^{(n)}). \end{cases}$$

We obtain

$$(2.12)\quad \begin{cases} a_{11}\left[\frac{1}{\omega}u_1^{(n+1)}+\left(1-\frac{1}{\omega}\right)u_1^{(n)}\right]+a_{12}u_2^{(n)}+a_{13}u_3^{(n)} = b_1 \\ a_{21}u_2^{(n+1)}+a_{22}\left[\frac{1}{\omega}u_2^{(n+1)}+\left(1-\frac{1}{\omega}\right)u_2^{(n)}\right]+a_{23}u_3^{(n)}=b_2 \\ a_{31}u_3^{(n+1)}+a_{32}u_2^{(n+1)}+a_{33}\left[\frac{1}{\omega}u_3^{(n+1)}+\left(1-\frac{1}{\omega}\right)u_3^{(n)}\right]=b_3. \end{cases}$$

Following the same procedure for the non-linear case we get

$$(2.13)\quad \begin{cases} f_1\left(\frac{1}{\omega}u_1^{(n+1)} + \left(1-\frac{1}{\omega}\right)u_1^{(n)}, u_2^{(n)}, u_3^{(n)}\right) = 0 \\ f_2\left(u_1^{(n+1)}, \frac{1}{\omega}u_2^{(n+1)}+\left(1-\frac{1}{\omega}\right)u_2^{(n)}, u_3^{(n)}\right) = 0 \\ f_3\left(u_1^{(n+1)}, u_2^{(n+1)}, \frac{1}{\omega}u_3^{(n+1)}+\left(1-\frac{1}{\omega}\right)u_3^{(n)}\right) = 0. \end{cases}$$

The symmetric successive overrelaxation method (SSOR method) (Sheldon [25]) is the same as the SOR method except that one computes $u_1^{(n+1/2)}$, $u_2^{(n+1/2)}$, $u_3^{(n+1/2)}$ based on $u_1^{(n)}$, $u_2^{(n)}$, $u_3^{(n)}$ using a <u>forward sweep</u>, and then computes $u_3^{(n+1)}$, $u_2^{(n+1)}$, $u_1^{(n+1)}$ based on $u_1^{(n+1/2)}$, $u_2^{(n+1/2)}$, $u_3^{(n+1/2)}$ using a <u>backward sweep</u>. The forward sweep is the same as the SOR method; thus the values $u_i^{(n+1/2)}$ are the same as one would get for $u_i^{(n+1)}$ using the SOR method. One then applies the SOR method using the equations in the reverse order to complete the iterative step.

3. Convergence Results for the Basic Methods.

We now discuss the convergence of the basic methods under various assumptions on the matrix A. We assume throughout that A is positive definite. Let us first consider the case where no additional assumptions are made on A.

Let us introduce the following notation. For any matrix A we let S(A) denote the <u>spectral radius</u> of A, i.e., the modulus of the eigenvalue of A of largest modulus. We define the norm of a matrix to be

$$||A|| = \sqrt{S(AA^H)}$$

where A^H is the complex conjugate transpose of A. If A has real eigenvalues, we let m(A) and M(A) denote, respectively, the smallest and largest eigenvalues of A.

Let us now consider the class of linear stationary iterative methods of the form

$$u^{(n+1)} = Gu^{(n)} + k \tag{3.1}$$

where I-G is nonsingular and

$$k = (I-G)A^{-1}b. \tag{3.2}$$

Such a method is completely consistent with (2.1) in the sense that the solution of the related equation

$$u = Gu + k \tag{3.3}$$

is a solution of (2.1) and vice versa. This implies, in particular, that if $u^{(n)} = A^{-1}b$ for some n, then

$u^{(n+1)} = u^{(n+2)} = \ldots = A^{-1}b$; moreover, if the sequence $u^{(0)}, u^{(1)}, \ldots$ defined by (3.1) converges, it converges to $A^{-1}b$.† We note that (3.1) can be derived from (2.1) by a <u>splitting</u> of A, i.e., a representation of A in the form

$$A = Q - R \tag{3.4}$$

where

$$Q = A(I-G)^{-1}, \quad R = A(I-G)^{-1}G. \tag{3.5}$$

Thus we can derive (3.1) by writing (2.1) in the form

$$Qu - Ru = b \tag{3.6}$$

and defining $u^{(n+1)}$ in terms of $u^{(n)}$ by

$$Qu^{(n+1)} - Ru^{(n)} = b \tag{3.7}$$

†We remark that the concept of complete consistency can be applied to linear stationary iterative methods for solving (2.1) where A may be singular. It can be shown [33][34] that (3.1) is completely consistent with (2.1) if and only if there exists a nonsingular matrix M such that $G = I - MA$ and $k = Mb$.

which, upon solving for $u^{(n+1)}$ yields (3.1).

In any case, the method (3.1) converges to $A^{-1}b$ for all $u^{(0)}$ if and only if $S(G) < 1$. The rapidity of convergence depends on the rapidity with which the norm $||G^n||$ tends to zero as $n \to \infty$. The number of iterations needed to reduce an initial error by a specified fraction is approximately inversely proportional to the _average rate of convergence_

$$R_n(G) = -\frac{1}{n} \log ||G^n||. \tag{3.8}$$

The _asymptotic average rate of convergence_ is given by

$$\lim_{n\to\infty} R_n(G) = -\log S(G). \tag{3.9}$$

In our subsequent discussion we shall refer to the average asymptotic rate of convergence simply as the "rate of convergence."

It can be shown (see, for instance, [29]) that the Gauss-Seidel method converges. Moreover, the SOR and the SSOR methods converge provided the relaxation factor ω lies in the range

$$0 < \omega < 2. \tag{3.10}$$

On the other hand, the Jacobi method need not converge. However, we can show that all of the eigenvalues μ of the corresponding matrix B are real and less than unity, i.e.,

$$m(B) \leq \mu \leq M(B) < 1. \tag{3.11}$$

Consequently, we can effectively use the <u>simultaneous overrelaxation method</u> defined by

$$\begin{aligned} u^{(n+1)} &= \omega(Bu^{(n)} + c) + (1-\omega)u^{(n)} \\ &= B_\omega u^{(n)} + \omega c \end{aligned} \tag{3.12}$$

where ω is the relaxation factor. This method is an <u>extrapolation</u> of the Jacobi method. It can be shown that the value of ω which minimizes $S(B_\omega)$ is given by

$$\omega_0 = \frac{2}{2-M(B)-m(B)} \tag{3.13}$$

and the corresponding spectral radius is

$$S(B_{\omega_0}) = \frac{M(B)-m(B)}{2-M(B)-m(B)} \text{.} \tag{3.14}$$

We refer to the method thus defined as the <u>optimum extrapolated Jacobi method</u> (J-OE method).

We now determine ω_0 and $S(B_{\omega_0})$ in terms of $m(\hat{A})$ and $M(\hat{A})$ where

$$\hat{A} = D^{-1/2}AD^{-1/2} \tag{3.15}$$

and D is the diagonal matrix with the same diagonal elements as A. Evidently we have

$$\begin{aligned} \hat{A} &= D^{-1/2}(D-C)D^{-1/2} \\ &= I-D^{-1/2}CD^{-1/2} \\ &= I-D^{1/2}BD^{-1/2} \end{aligned} \tag{3.16}$$

since $B = D^{-1}C$. Since $D^{1/2}BD^{-1/2}$ is similar to B it follows that μ is an eigenvalue of B if and only if $\nu = 1-\mu$ is an eigenvalue of $\hat{A}$. Since A, and hence $\hat{A}$, is positive definite, it follows that each eigenvalue ν of $\hat{A}$ is positive and hence each eigenvalue μ of B is less than unity. Moreover, we have

$$(3.17)\qquad \begin{cases} m(\hat{A}) = 1 - M(B) \\ M(\hat{A}) = 1 - m(B) \end{cases}$$

and hence

$$(3.18)\qquad \omega_0 = \frac{2}{M(\hat{A})+m(\hat{A})}$$

and

$$(3.19)\qquad S(B_{\omega_0}) = \frac{M(\hat{A})-m(\hat{A})}{M(\hat{A})+m(\hat{A})}$$

$$= \frac{K(\hat{A})-1}{K(\hat{A})+1}$$

where $K(\hat{A})$ is the condition[†] of $\hat{A}$ defined by

$$(3.20)\qquad K(\hat{A}) = \frac{M(\hat{A})}{m(\hat{A})}.$$

The quantity $R(B_{\omega_0})$ given by

$$(3.21)\qquad R(B_{\omega_0}) = -\log S(B_{\omega_0})$$

[†]In general one can define the condition of a nonsingular matrix A by $K(A) = ||A||\cdot||A^{-1}||$. However, if A is positive definite then $K(A) = M(A)/m(A)$.

is our standard of convergence, and will be used as a basis with which to compare other methods.

An alternative standard of convergence could be defined in terms of the optimum extrapolated RF method (RF-OE method). The RF method* is defined by

$$(3.22) \qquad u^{(n+1)} = u^{(n)} + p(Au^{(n)}-b)$$

where p is a nonvanishing constant. The extrapolated RF method is defined by

$$(3.23) \qquad u^{(n+1)} = \omega\left[u^{(n)}+p\{Au^{(n)}-b\}\right] + (1-\omega)u^{(n)}.$$

The optimum value of ω is given by

$$(3.24) \qquad \omega_0 = -\frac{(2/p)}{m(A)+M(A)}$$

and the corresponding spectral radius is

$$(3.25) \qquad S((R_p)_{\omega_0}) = \frac{M(A)-m(A)}{M(A)+m(A)} = \frac{K(A)-1}{K(A)+1}$$

where

---

*The RF method is the method of Richardson [24] with a fixed iteration parameter.

(3.26) $$K(A) = \frac{M(A)}{m(A)} .$$

Thus the spectral radius of the RF-OE method is independent of p and is the same as that of the J-OE method except that $K(\hat{A})$ is replaced by $K(A)$. We prefer to consider the J-OE method rather than the RF-OE method as a standard of comparison because of its invariance under the transformation $A \to EAE$ where E is a diagonal matrix with positive diagonal elements. Another reason for preferring the J-OE method is that, from a result of Forsythe and Strauss, it follows that if A has Property A then $K(\hat{A}) \leq K(A)$. Thus the J-OE method is at least as good as the RF-OE method in this case.

It can be shown that in certain cases the (average asymptotic) rate of convergence of the SOR method is greater by an order-of-magnitude than that of the J-OE method. For such cases

(3.27) $$R(\mathcal{L}_{\omega_b}) \sim 2\sqrt{2}\,\sqrt{R(B_{\omega_0})}$$

asymptotically as $R(B_{\omega_0}) \to 0$, where $\mathcal{L}_{\omega_b}$ refers to the matrix corresponding to the SOR method and where

$$(3.28) \qquad \omega_b = \frac{2}{1 + \sqrt{1-\bar{\mu}^2}}$$

and

$$(3.29) \qquad \bar{\mu} = S(B).$$

The above results hold if A is a <u>consistently ordered</u> matrix. For the precise definition of consistently ordered matrices we refer to [33]. We simply note here that, given a matrix with Property A, for certain permutation matrices P, $P^{-1}AP$ is consistently ordered. A matrix has Property A if, for some permutation matrix P, the matrix $P^{-1}AP$ has the form*

$$(3.30) \qquad P^{-1}AP = \begin{pmatrix} D_1 & H \\ K & D_2 \end{pmatrix}$$

where D_1 and D_2 are square diagonal matrices.

---

*For this definition we assume that the order of the matrix, N, is greater than one.

We remark that the matrix corresponding to a five-point difference equation such as that derived in Section 1 has Property A. Moreover, if the ordering of the rows and columns corresponds to the "red-black ordering" then A has the form (3.30). If A corresponds to the "natural ordering," where one proceeds from left to right and up, the matrix does not in general have the form (3.30) but is nevertheless consistently ordered.

If A is consistently ordered, then $S(\mathcal{L}_{\omega_b}) \leq S(\mathcal{L}_{\omega})$ for any $\omega \neq \omega_b$. Moreover,

$$S(\mathcal{L}_{\omega_b}) = \omega_b - 1. \tag{3.31}$$

→ If the matrix A has the form

$$A = \begin{pmatrix} D_1 & H \\ K & D_2 \end{pmatrix} \tag{3.32}$$

where D_1 and D_2 are square diagonal matrices, then one can define the <u>modified successive overrelaxation method</u> (MSOR method) as follows [9][37]. We write the system (2.1) in the form

(3.33)
$$\begin{cases} D_1u_1 + H\,u_2 = b_1 \\ K\,u_1 + D_2u_2 = b_2 \end{cases}$$

where u_1, u_2, b_1, b_2 are column vectors of appropriate sizes. The MSOR method is defined by

(3.34)
$$\begin{cases} u_1^{(n+1)} = \omega_{n+1}(Fu_2^{(n)}+c_1) + (1-\omega_{n+1})u_1^{(n)} \\ u_2^{(n+1)} = \omega'_{n+1}(Gu_1^{(n)}+c_2) + (1-\omega'_{n+1})u_2^{(n)} \end{cases}$$

where

(3.35)
$$\begin{cases} F = -D_1^{-1}H, \quad G = -D_2^{-1}K \\ c_1 = D_1^{-1}b_1, \quad c_2 = D_2^{-1}b_2. \end{cases}$$

If $\omega_n = \omega'_n = \omega$ for all ω, then we have the ordinary SOR method. It can be shown [37] that the spectral radius of the matrix corresponding to the MSOR method is minimized by letting $\omega_1 = \omega'_1 = \ldots = \omega_b$. However, the _cyclic Chebyshev semi-iterative method_ (CCSI method) yields a set of matrices with smaller norms. (See [15][16].) The relaxation factors corresponding to the CCSI method are

(3.36)
$$\begin{cases} \omega_1 = 1, \qquad \omega_2 = \dfrac{2}{2-\bar{\mu}^2} \\ \omega_k = [1-\frac{1}{4}\omega'_{k-1}\bar{\mu}^2]^{-1}, \omega'_k=[1-\frac{1}{4}\omega_k\bar{\mu}^2]^{-1}, \end{cases}$$

$$k = 2,3,\ldots$$

Even though A is not consistently ordered, the SOR theory holds, with minor changes, for an important class of matrices, namely, the class of matrices such that

(3.37)
$$\begin{cases} a_{i,i} > 0 \\ a_{i,j} \leq 0, \qquad j \neq i. \end{cases}$$

Such matrices are referred to as "L-matrices" by Young [33]. We remark that a positive definite L-matrix is a Stieltjes matrix.

Kahan [18] showed that if A is a Stieltjes matrix and if ω_b is given by (3.28), then

(3.38)
$$\omega_b - 1 \leq S(\mathcal{L}_{\omega_b}) \leq \sqrt{\omega_b - 1} \ .$$

Let us now compare the SOR method with the J-OE method. In the case A has Property A, we know that $-m(B) = M(B) = \bar{\mu}$. In the case where A is a Stieltjes matrix, we can only say that $m(B) \leq 0^*$ and, by the Perron-Frobenius theory, $-m(B) \leq M(B) = \bar{\mu}$. However, by (3.14), we can still assert that

$$S(B_{\omega_0}) \geq \frac{\bar{\mu}}{2-\bar{\mu}} . \tag{3.39}$$

This follows since the function (of x)

$$\frac{M(B)-x}{2-M(B)-x} \tag{3.40}$$

where $0 \leq M(B) < 1$ is minimized in the range $0 \geq x \geq -M(B)$ when $x = 0$. Therefore we have, asymptotically as $\bar{\mu} \to 1-$,

$$\begin{cases} R(B_{\omega_0}) \underset{\sim}{\leq} 2R(B) \\ \\ R(\mathcal{L}_{\omega_b}) \underset{\sim}{\geq} \sqrt{R(B_{\omega_0})/2}. \end{cases} \tag{3.41}$$

---

*This follows since all diagonal elements of B vanish. Hence the sum of the eigenvalues of B vanishes. Thus at least one eigenvalue of B must be nonpositive.

Thus, the advantage of the SOR method over the J-OE method may be as little as $\frac{1}{4}$ what it was in the consistently ordered case. On the other hand, we still have an order of magnitude improvement in the convergence.

While neither of the conditions we have so far considered, namely, that A be consistently ordered or that A be an L-matrix, is <u>necessary</u> for the SOR method to be effective, the following example shows that <u>some</u> condition is required. Thus, if $0 < a < 1$, the matrix

$$(3.42) \qquad A = \begin{pmatrix} 1 & a & a \\ a & 1 & a \\ a & a & 1 \end{pmatrix}$$

is positive definite but is neither consistently ordered nor an L-matrix. Moreover, the SOR method is not appreciably more effective than the J-OE method. (See, for instance, [33, Ch. 12].) On the other hand, we shall show in Section 4 that methods which are better by an order-of-magnitude than the J-OE method can easily be constructed.

We now introduce another class of matrices which is not entirely included in the classes previously considered. We consider the case where

$$S(LU) \leq \frac{1}{4} \tag{3.43}$$

where L and U are strictly lower and strictly upper triangular matrices, respectively, such that

$$L + U = B. \tag{3.44}$$

(Here B is the matrix of the Jacobi method.) For the model problem the condition is satisfied provided the natural ordering is used. (It is <u>not</u> satisfied if the red-black ordering is used.) The condition is also satisfied for all regions -- not just for the square or rectangle.

Using a slight extension of the results of Young [33][35] one can show that if we let

$$\omega^* = \frac{2}{1 + \sqrt{2(1-M(B))}} \tag{3.45}$$

then we have

(3.46) $$S(\mathcal{L}_{\omega^*}) \leq \sqrt{\left[\frac{1 - \sqrt{\frac{1-M(B)}{2}}}{1 + \sqrt{\frac{1-M(B)}{2}}}\right]}$$

From (3.14) and the fact that $m(B) \leq 0$ it follows as in the derivation of (3.9)

(3.47) $$S(B_{\omega_0}) \geq \frac{M(B)}{2-M(B)} .$$

Hence, asymptotically as $M(B) \to 1-$ we have

(3.48) $$R(B_{\omega_0}) \lesssim 2(-\log M(B)).$$

Consequently,

(3.49) $$R(\mathcal{L}_{\omega^*}) \gtrsim \frac{1}{2}\sqrt{R(B_{\omega_0})} .$$

It can also be shown* that for the SSOR

*See [33][35]. For earlier references on the SSOR method see the work of Habetler and Wachspress [17] and Ehrlich [13][14].

$$(3.50) \qquad S(\mathscr{S}_{\omega^*}) \leq \frac{1 - \sqrt{\frac{1-M(B)}{2}}}{1 + \sqrt{\frac{1-M(B)}{2}}}$$

and, asymptotically,

$$(3.51) \qquad R(\mathscr{S}_{\omega^*}) \gtrsim \sqrt{R(B_{\omega_0})}.$$

Here $\mathscr{S}_{\omega^*}$ is the matrix corresponding to the SSOR method with $\omega = \omega^*$. Evidently the SSOR method is better, by an order-of-magnitude than the J-OE method.

We remark that if A is also consistently ordered (or even if A has Property A), then $B_{\omega_0} = B$ and we have

$$(3.52) \qquad R(\mathscr{S}_{\omega^*}) \gtrsim \sqrt{2}\ \sqrt{R(B_{\omega_0})}$$

while

$$R(\mathscr{L}_{\omega_b}) \gtrsim 2\sqrt{2}\ \sqrt{R(B_{\omega_0})}.$$

Thus, the SSOR method is only about half as fast as the SOR method even though it requires twice as much work per iteration. However, the eigenvalues of

$\mathcal{L}_{\omega^*}$ are real and positive, and as we show in the next section, the convergence can be accelerated by another order-of-magnitude using a semi-iterative method.

4. Semi-iterative Methods and Second-degree Methods.

We now consider the acceleration of the convergence of the linear stationary iterative method (3.42) when the eigenvalues of G are real and less than unity. (We do not assume that $S(G) < 1$.) For each set of real numbers

$$\begin{array}{llll} \alpha_{00} & & & \\ \alpha_{10} & \alpha_{11} & & \\ \alpha_{20} & \alpha_{21} & \alpha_{22} & \\ \alpha_{30} & \alpha_{31} & \alpha_{32} & \alpha_{33} \\ & \cdots & \cdots & \end{array}$$

such that

$$\sum_{k=0}^{n} \alpha_{n,k} = 1, \qquad n = 0,1,2,\ldots \tag{4.1}$$

we define the sequence

$$(4.2) \qquad v^{(n)} = \sum_{k=0}^{n} \alpha_{n,k} u^{(k)}, \qquad n = 0,1,2,\ldots$$

where $u^{(n)}$ are generated from (3.2) using the basic method. Such an acceleration procedure is called a <u>semi-iterative method</u> by Varga [27] (See also [5][15] and [16].). It can be shown that the optimum values of the $\alpha_{n,k}$ can be obtained in terms of m(G) and M(G) based on the use of Chebyshev polynomials. It can also be shown that for the optimum method, one can express $v^{(n+1)}$ in terms of $v^{(n)}$ and $v^{(n-1)}$ -- it is not necessary to first compute the $u^{(n)}$ -- or for that matter the $\alpha_{n,k}$. Thus we have, with $v^{(0)}$ arbitrary

$$(4.3) \qquad v^{(n+1)} = \omega_{n+1}[\omega_0[Gv^{(n)}+k]+(1-\omega_0)v^{(n)}] + (1-\omega_{n+1})v^{(n-1)}$$

where

$$(4.4) \qquad \omega_0 = \frac{2}{2-M(G)-m(G)}$$

$$
(4.5) \qquad \begin{cases} \omega_1 = 1 \\ \omega_2 = \dfrac{2z^2}{2z^2-1} \\ \omega_{n+1} = \left[1 - \dfrac{1}{4z^2}\,\omega_n\right]^{-1}, \quad n = 1,2,\ldots \end{cases}
$$

and where

$$
(4.6) \qquad z = \frac{2-M(G)-m(G)}{M(G)-m(G)} .
$$

Let $P_n(x)$ be defined by

$$
(4.7) \qquad P_n(x) = \sum_{k=0}^{n} \alpha_{n,k} x^k, \quad n = 0,1,\ldots .
$$

Then one can show that

$$
(4.8) \qquad S(P_n(G)) \le \frac{2\tau^n}{1+\tau^{2n}}
$$

where

$$
(4.9) \qquad \begin{cases} \tau = \dfrac{\sigma}{1+\sqrt{1-\sigma^2}} \\ \sigma = \dfrac{1}{z} . \end{cases}
$$

We note that σ is the spectral radius of the

optimum extrapolated method based on the original method -- see the discussion of Section 3 (especially (3.13) and (3.14)) for the Jacobi method. The asymptotic average rate of convergence is

$$\lim_{n\to\infty} \frac{1}{n} (-\log S(P_n(G)) = -\log \tau \tag{4.10}$$

$$\sim \sqrt{2}\ \sqrt{-\log \sigma}$$

as $\sigma \to 1$. Thus, the asymptotic average rate of convergence of the optimum accelerated method corresponding to a given method is better by an order-of-magnitude than that of the optimum extrapolated method.

Applying the above analysis to the case of the Jacobi method we see that even if nothing is assumed about A except that it is positive definite we can still improve on the J-OE method by an order-of-magnitude using the optimum semi-iterative method based on the Jacobi method (the J-SI method).

If the matrix A is consistently ordered, then the J-SI method converges approximately half as fast as the SOR method. For in this case $m(B) = -M(B) = \bar{\mu}$

and

$$(4.11) \qquad \tau = \sqrt{\omega_b - 1}.$$

Hence

$$(4.12) \qquad R(\mathcal{L}_{\omega_b}) \sim 2R(P_n(B)).$$

On the other hand, if the matrix A has the form (3.32), then, as shown by Golub and Varga [15][16], one can "compress" the iterative process -- in other words, one can only change half of the values on each iteration. The resulting method is precisely the CCSI method which has been discussed in the previous section.

A second application of semi-iterative methods* is for the acceleration of the convergence of the SSOR method in the case where A satisfies the condition (3.43). Two order-of-magnitude improvements can be made: the first in going from the J-OE

---

*We remark that semi-iterative methods are not useful for the SOR method since the eigenvalues are complex. (See the discussion of Varga [27].)

method to the SSOR method; second in going from the SSOR method to the SSOR-SI-method. The eigenvalues of the matrix $\mathcal{S}_{\omega^*}$ are real and positive. It can be shown that the rate of convergence of the SSOR-SI method is, asymptotically as $R(B_{\omega_0}) \to 1-$,

$$(4.13) \qquad R \gtrsim 2[R(B_{\omega_0})]^{1/4}.$$

If the matrix A is also consistently ordered (actually Property A would be sufficient), then for the SSOR method we have*

$$(4.14) \qquad R \gtrsim 2^{5/4}[R(B_{\omega_0})]^{1/4}.$$

The situation is summarized in Table 4.1. The rates of convergence for the methods indicated

---

*The constants appearing in (4.13) and (4.14) are derived as follows. Since the eigenvalues of $\mathcal{S}_{\omega^*}$ are real and positive, the optimum extrapolated SSOR method (SSOR-OE method) converges roughly <u>twice</u> as fast as the SSOR method. The semi-iterated method has a rate of convergence approximately $\sqrt{2}$ times the square root of the rate of convergence of the optimum extrapolated method.

are given under various assumptions on the matrix A. The results are asymptotic as $R(B_{\omega_0}) \to 1-$.

Based on the above results we offer the following conclusions:

(1) If nothing** is known about A except that it is positive definite, then the J-SI method should be used.

(2) If A is consistently ordered, one can either use the SOR method or else permute the rows and columns of A, if necessary, to obtain the form (3.32), whereupon the CCSI method can be used. Another possibility is to perform one iteration with $\omega = 1$ followed by using the SOR method with $\omega = \omega_b$.

(3) If A is a Stieltjes matrix, one can either use the SOR method or the J-SI method. Because less storage is

---

**Of course we must also assume that good bounds for the eigenvalues of $\hat{A}$ are available.

required for the SOR method and also because it may converge faster than $\frac{\sqrt{2}}{2}\sqrt{R(B_{\omega_0})}$, I would generally prefer it to the J-SI method.

(4) If $S(LU) \le 1/4$, the SSOR-SI method should be used.

We remark that (4.3) defines a <u>second-degree</u> method (since $v^{(n+1)}$ is determined from $v^{(n)}$ and $v^{(n-1)}$). (The basic method (3.42) is a <u>first-degree</u> method since $u^{(n+1)}$ depends only on $u^{(n)}$.) It is shown by Golub and Varga [15][16] and by Young [33] [35] that one can obtain nearly as rapid convergence as for the optimum semi-iterative method by letting $\omega_2, \omega_3, \ldots$ be equal to the limiting value, namely $\hat{\omega}_b$ given by

$$\hat{\omega}_b = \frac{2}{1+\sqrt{1-\sigma^2}}$$

where σ is given by (4.9). That $\hat{\omega}_b$ is the limiting value of the ω_n follows from (4.5). It can also be shown [36] that for the first iteration one can use

TABLE 4.1 RATES OF CONVERGENCE OF VARIOUS METHODS

Method	No Additional Assumptions	Consistently Ordered	Stieltjes Matrix	$S(LU) \le \frac{1}{4}$	Consistently Ordered and $S(LU) \le 1/4$
SOR	*	$2\sqrt{2}\ \sqrt{R(B_{\omega_0})}$	$\sqrt{R(B_{\omega_0})/2}$	$1/2\sqrt{R(B_{\omega_0})}$	$2\sqrt{2}\ \sqrt{R(B_{\omega_0})}$
CCSI	*	$2\sqrt{2}\ \sqrt{R(B_{\omega_0})}$ **	*	*	$2\sqrt{2}\ \sqrt{R(B_{\omega_0})}$ **
SSQR	*	*	*	$\sqrt{R(B_{\omega_0})}$	$\sqrt{2}\ \sqrt{R(B_{\omega_0})}$
J-SI	$\sqrt{2}\ \sqrt{R(B_{\omega_0})}$	$\sqrt{2}\ \sqrt{R(B_{\omega_0})}$	$\sqrt{2}\ \sqrt{R(B_{\omega_0})}$	$\sqrt{2}\ \sqrt{R(B_{\omega_0})}$	$\sqrt{2}\ \sqrt{R(B_{\omega_0})}$
SSOR-SI	*	*	*	$2[R(B_{\omega_0})]^{1/4}$	$2^{5/4}[R(B_{\omega_0})]^{1/4}$

*May not be effective.

**Assumed to have the form $A = \begin{pmatrix} D_1 & H \\ K & D_2 \end{pmatrix}$.

the slightly more general formula

$$v^{(1)} = \omega_1'(Gv^{(0)}+k) + (1-\omega_1')v^{(0)}$$

and the choice of $\omega = \hat{\omega}_b$ will, in a sense, give the optimum convergence rate for any stationary second-degree method of the form

$$v^{(n+1)} = \omega\{\omega_0[Gv^{(n)}+k]+(1-\omega_0)v^{(n)}\}+(1-\omega)v^{(n-1)},$$

$$n = 1,2,\ldots .$$

This is true for any choice of ω_1'.

5. Application to the Solution of Elliptic Boundary Value Problems*.

We now consider the generalized Dirichlet problem involving the self-adjoint elliptic equation

$$\frac{\partial}{\partial x}\left(A\frac{\partial u}{\partial x}\right) + \frac{\partial}{\partial y}\left(C\frac{\partial u}{\partial y}\right) + Fu = G \tag{5.1}$$

---

*For an up-to-date account of methods for solving elliptic boundary value problems the reader is referred to the book by Birkhoff [2].

where $A > 0$, $C > 0$, and $F \leq 0$ in the closure of the region under consideration. As in the case of the model problem we cover the region with a square mesh of size h. We assume that for a sequence of values of h tending to zero the region is regular in the sense that, for any mesh point in the interior of the region, the four neighboring mesh points are in the region or on the boundary.

We consider the use of a five-point difference equation** based on the use of the difference representation

$$\frac{\partial}{\partial x}\left(A \frac{\partial u}{\partial x}\right) \sim h^{-2}\left\{A(x+\frac{h}{2},y)[u(x+h,y)-u(x,y)] - A(x-\frac{h}{2},y)[u(x,y)-u(x-h,y)]\right\} \tag{5.2}$$

and a similar expression for $\frac{\partial}{\partial x}\left(C \frac{\partial u}{\partial y}\right)$. Substituting in the differential equation and multiplying by $-h^2$ leads to the linear system

---

**We remark that, for the Laplace equation, if one uses the usual 9-point formula then A does not have Property A. However, A is a Stieltjes matrix. Moreover, if the natural ordering is used, $S(LU) \leq 1/4$.

$$Au = b \tag{5.3}$$

where A is a positive definite matrix with Property A. Moreover, if the mesh points are labelled in the "natural ordering" or the "red-black ordering," then A is consistently ordered.

A bound for M(B) (and also m(B), which equals -M(B)) can be given [32][33]. Thus we have, for the case $F \equiv 0$,

$$M(B) \le \tag{5.4}$$

$$1 - \frac{2\underline{A}\sin^2\frac{\pi}{2I} + 2\underline{C}\sin^2\frac{\pi}{2J}}{\frac{1}{2}(\overline{A}+\underline{A}) + \frac{1}{2}(\overline{C}+\underline{C}) + \frac{1}{2}(\overline{A}-\underline{A})\cos\frac{\pi}{I} + \frac{1}{2}(\overline{C}-\underline{C})\cos\frac{\pi}{J}}$$

where the region is included in an $Ih \times Jh$ rectangle and where

$$\begin{aligned} &\underline{A} \le A(x,y) \le \overline{A} \\ &\underline{C} \le C(x,y) \le \overline{C} \end{aligned} \tag{5.5}$$

For the case of Laplace's equation in the

square (the model problem), we have the following asymptotic results.

Method	Rate of Convergence
Jacobi (same as J-OE)	$\frac{\pi^2}{2} h^2$
Gauss-Seidel	$\pi^2 h^2$
SOR	$2\pi h$
CCSI (red-black ordering)	$2\pi h$
SSOR (natural ordering)	πh
J-SI	πh
SSOR-SI (natural ordering)	$2\sqrt{\pi}\ h^{1/2}$

The question as to what assumptions on $A(x,y)$ and $C(x,y)$ are needed so that $S(LU) \leq 1/4$ is not settled. It can be shown [33] that with the SSOR-SI method the reciprocal rate of convergence is $O(h^{-3/4})$. Phein [22] showed that one obtains $O(h^{-1/2})$ in the case of the differential equation

$$(5.6) \qquad \frac{\partial}{\partial x}\left(\frac{1}{y}\frac{\partial u}{\partial x}\right) + \frac{\partial}{\partial y}\left(\frac{1}{y}\frac{\partial u}{\partial y}\right) = 0.$$

Further theoretical and experimental work on this question is underway.

Numerical Results

The following table summarizes the numerical results* obtained for the model problem with mesh size h. The number of iterations, n, required to attain a certain level of convergence is given in each case.

	SOR		SSOR-SI		SSOR-SD		ADI	
h^{-1}	ω	n	ω	n	ω	n	ω	n
20	1.74	53	1.76	17	1.73	18	4	11
40	1.86	117	1.87	24	1.85	27	5	14
80	1.94	236	1.94	34	1.92	39	5	18

The SSOR-SD method is the stationary second-degree method based on the SSOR method. The results for the

---

*The numerical results were obtained using facilities of The University of Texas Computation Center.

ADI method refer to the alternating direction method of Peaceman and Rachford [21], which is discussed in Section 6. The number of different iteration parameters is denoted by m.

It can be seen that the SSOR-SD method is nearly as good as the SSOR-SI method. Both are substantially better than the SOR method -- note, however, that twice as much work per iteration and more storage is required using the SSOR methods. The ADI method is considerably faster than the other methods and the advantage would increase as h decreases.

6. Other Methods.

We have by no means covered all of the methods which are currently being used for solving large linear systems with sparse matrices. One important omission is the class of alternating direction implicit methods. Such methods are very powerful in case A can be split into the sum

$$A = H + V \tag{6.1}$$

where H and V are positive definite matrices such that

$$(6.2)\qquad HV = VH,$$

and such that for any $\rho > 0$ one can easily solve any linear system involving the matrix $H + \rho I$ or $V + \rho I$. Unfortunately, as shown by Birkhoff and Varga [3] and by Birkhoff, Varga, and Young [4] the conditions are satisfied exactly in very rare cases. If A is derived from a self-adjoint elliptic equation, the region must be a rectangle and the equation must be separable. However, Widlund [30][31] has modified the method to extend the range of its applicability.

In the case of the model problem, the reciprocal rate of convergence is $O(|\log h|)$ -- thus the method is much faster than the other methods considered above. Moreover, numerical experiments for certain cases where the theory does not hold indicate that the method works quite well [4]. Nevertheless, there are cases, involving highly varying coefficients, where the method fails to converge

[23].

Possibly in part due to the lack of theoretical results concerning the method, its use in actual problems is much less than might be expected on the basis of its spectacular performance for the model problem.

Another promising class of methods are those based on "quasi-direct" procedures. In the case of the model problem, or more generally for a differential equation with constant coefficients in a rectangle, one can represent the solution in terms of a finite Fourier series. By various ingenious devices, this solution can be evaluated very rapidly [6][10].

Recently, Concus and Golub [7] have extended the method to apply to an equation of the form (5.1) with $A(x,y) \equiv C(x,y)$. Here a direct method is used in combination with an iterative method.

The idea of using direct methods in combination with iterative methods is by no means new. Indeed, block, or line, methods are frequently used. Generally one can obtain a modest factor of improvement, of the order of 2 or so, using block methods. For references on block methods the reader is referred to [1],[8] and [28].

Another class of methods is the class of "almost factorization methods" (See the papers of Stone [26], DuPont [11], and DuPont, Kendall, and Rachford [12]). For the case of the model problem, one can obtain convergence rate comparable to that of the SSOR-SI method, using these methods.

REFERENCES

[1] R. F. Arms, L. D. Gates, and B. Zondek, A method of block iteration, J. Soc. Indust. Appl. Math. 4 (1956) 220-229.

[2] G. Birkhoff, The Numerical Solution of Elliptic Equations, CBMS Regional Conference Series in Applied Mathematics, Vol. 1, SIAM, Philadelphia, Pa., 1971.

[3] G. Birkhoff and R. S. Varga, Implicit alternating direction methods, Trans. Amer. Math. Soc., 92 (1959) 13-24.

[4] G. Birkhoff, R. S. Varga and D. Young, Alternating direction implicit methods, Advances in Computers, Vol. 3, F. Alt and M. Rubinoff, eds., Academic Press (1962) 189-273.

[5] A. Blair, N. Metropolis, J. von Neumann, A. H. Taub, and M. Tsingori, A study of a numerical solution of a two-dimensional hydrodynamical problem, Math. Tables Aids Comput., 13 (1959) 145-184.

[6] B. L. Buzbee, F. W. Dorr, J. A. George, and G. H. Golub, The direct solution of the discrete Poisson equation on irregular regions, SIAM J. Numer. Anal. 8 (1971) 722-736.

[7] P. Concus and G. Golub, Use of fast direct methods for the effective numerical solution of nonseparable elliptic equations, unpublished manuscript, (1972).

[8] E. H. Cuthill and R. S. Varga, A method of normalized block iteration, J. Assoc. Comput. Mach. 6 (1959) 236-244.

[9] R. DeVogelaere, Over-relaxations, Abstract No. 539-53, Amer. Math. Soc. Notices 5 (1958) 147.

[10] F. W. Dorr, The direct solution of the discrete Poisson equation on a rectangle, SIAM Review 12 (1970) 248-263.

[11] T. Dupont, A factorization procedure for the solution of elliptic difference equations, SIAM Jour. Numer. Anal., 5 (1968) 753-782.

[12] T. Dupont, R. P. Kendall, and H. H. Rachford, Jr., An approximate factorization procedure for solving self-adjoint elliptic difference equations, SIAM Jour. Numer. Anal., 5 (1968) 559-573.

[13] L. W. Ehrlich, The block symmetric successive overrelaxation method, Doctoral thesis, Univ. of Texas, Austin, 1963.

[14] L. W. Ehrlich, The block symmetric successive overrelaxation method, J. Soc. Indust. Appl. Math., 12 (1964) 807-826.

[15] G. H. Golub and R. S. Varga, Chebyshev semi-iterative methods, successive over-relaxation iterative methods, and second-order Richardson methods, Part I, Numer. Math., 3 (1961) 147-156.

[16] G. H. Golub and R. S. Varga, Chebyshev semi-iterative methods, successive over-relaxation iterative methods, and second-order Richardson iterative methods, Part II, Numer. Math., 3 (1961) 157-168.

[17] G. J. Habetler and E. L. Wachspress, Symmetric successive overrelaxation in solving difference equations, Math. Comp., 15 (1961) 356-362.

[18] W. Kahan, Gauss-Seidel methods of solving large systems of linear equations, Doctoral thesis, University of Toronto, 1958.

[19] N. K. Nichols, On the convergence of two-stage iterative processes for solving linear equations, SIAM J. Numer. Anal., to appear.

[20] J. M. Ortega and W. C. Rheinboldt, Iterative Solution of Nonlinear Equations in Several Variables, Academic Press, New York, 1970.

[21] D. W. Peaceman and H. H. Rachford, Jr., The numerical solution of parabolic and elliptic differential equations, J. Soc. Indus. Appl. Math., 3 (1955) 28-41.

[22] T. Phein, An application of semi-iterative and second-degree symmetric successive overrelaxation iterative methods, Master's thesis, University of Texas at Austin, 1972.

[23] H. Price and R. S. Varga, Recent Numerical experiments comparing successive overrelaxation iterative methods with implicit alternating direction methods, Report No. 91, Gulf Research and Development Co., Pittsburgh, Pa., 1962.

[24] L. F. Richardson, The approximate arithmetical solution by finite differences of physical problems involving differential equations with an application to the stresses in a masonry dam, Philos. Trans, Roy. Soc. London Ser. A, 210 (1910) 307-357.

[25] J. Sheldon, On the numerical solution of elliptic difference equations, Math. Tables Aids Comput., 9 (1955) 101-112.

[26] H. L. Stone, Iterative solution of implicit approximations of multidimensional partial differential equations, SIAM J. Numer. Anal., 5 (1968) 530-558.

[27] R. S. Varga, A comparison of the successive overrelaxation method and semi-iterative methods using Chebyshev polynomials, Jour. Soc. Indus. Appl. Math., 5 (1957) 39-46.

[28] R. S. Varga, Factorization and normalized iterative methods, Boundary Problems in Differential Equations, R. E. Langer, ed., Univ. of Wisconsin Press, Madison (1960) 121-142.

[29] R. S. Varga, Matrix Iterative Analysis, Prentice-Hall, Inc., Englewood Cliffs, New Jersey, 1962.

[30] O. B. Widlund, On the effects of scaling of the Peaceman-Rachford method, Conference on the Numerical Solution of Differential Equations, J. L. Morris, ed., Lecture Notes in Mathematics, Vol. 109, Springer-Verlag, Berlin, 1969, 113-132.

[31] O. B. Widlund, On the rate of convergence of an alternating direction implicit method in a noncommutative case, Math. Comp., 20 (1966) 500-515.

[32] D. M. Young, A bound for the optimum relaxation factor for the successive overrelaxation method, Numer. Math., 16 (1971) 408-413.

[33] D. M. Young, Iterative Solution of Large Linear Systems, Academic Press, New York, 1971.

[34] D. M. Young, On the consistency of linear stationary iterative methods, SIAM J. Numer. Anal., 9, (1972), 89-96.

[35] D. M. Young, Second-degree iterative methods for the solution of large linear systems, J. Approx. Theory, 5 (1972) 137-148.

[36] D. M. Young and D. R. Kincaid, Linear stationary second-degree methods for the solution of large linear systems, in preparation.

[37] D. M. Young, M. F. Wheeler and J. A. Downing, On the use of the modified successive overrelaxation method with several relaxation factors, Proceedings of IFIP Congress 65 [Information Processing 1965], Vol. I, Spartan, Washington, D. C., 1965, 177-182.

SOME COMPUTATIONAL TECHNIQUES FOR THE NONLINEAR LEAST SQUARES PROBLEM

J. E. Dennis, Jr.*

1. Introduction.

Let F be a transformation of E^N into E^M, where N and M are integers and where E^I denotes the I dimensional space of real column vectors, for any integer I. This paper will be concerned with the solving of the equation $F(x) = 0 \in E^M$. In event that $N = M$, we generally try to satisfy the equation exactly. However, it is often the case that $N < M$ and even the most optimistic analyst would admit the necessity for relaxing the notion of a solution for such an overdetermined problem.

We shall adopt the convention that x* solves the overdetermined system of nonlinear equations,

---

*This work was supported by NSF Grant GJ-27528.

$F(x) = 0$, when x^* minimizes the functional

$$\phi(x) = (1/2)F(x)^T F(x),$$

although $\phi_W(x) = (1/2)F(x)^T WF(x)$ is often used. Here, W is some given positive definite symmetric matrix of order M, which is chosen through statistical or other considerations. (cf. [22] and [23] for an interpretation of W as an error covariance matrix.)

Our purpose here will be to present one way of looking at some of the more useful computational techniques for solving the overdetermined nonlinear least squares problem. In particular, we will show that these methods can profitably be viewed as Newton-like methods applied to the solution of $\nabla\phi(x) = 0$. No attempt will be made to survey all available methods, but we do hope to provide the reader with a framework within which to view some more esoteric methods.

2. The Methods in General.

Let us assume the notation of §1 and proceed

to consider the minimization of

$$\phi(x) = (1/2)F(x)^T F(x). \tag{1}$$

If F is twice differentiable, then it is easy to show that

$$\nabla\phi(x) = J_F(x)^T F(x) \tag{2}$$

where $J_F(x) = [(\partial f_i/\partial x_j)|_x]$ is the Jacobian matrix of $F = (f_1, f_2, \ldots f_M)^T$ evaluated at x and that the Hessian[#] of ϕ is

$$\nabla^2\phi(x) = \sum_{i=1}^{M} f_i(x)\nabla^2 f_i(x) + J_F(x)^T J_F(x). \tag{3}$$

The special forms of (2) and (3) show an interesting thing. When one has computed J_F, this not only furnishes $\nabla\phi$, but also a part of $\nabla^2\phi$. Hence, some of the methods for the general unconstrained minimization problem, such as the quasi-Newton or modification methods, are suspect for this special

---

[#] Here, ∇^2 is not the Laplace operator. Rather, $[\nabla^2\phi]_{ij} = \partial^2\phi/\partial x_i \partial x_j$ throughout this paper.

case. Their forte is in building up a Hessian approximation from gradient information alone. If one applies such a method to $\nabla\phi$ in a straightforward way, then the term $J_F(x)^T J_F(x)$ would be thrown away -- the very antithesis of the philosophy of these methods. In fact, there is evidence to indicate that practical experience follows aesthetics in this case [9, Tables A and B].

All of the methods here have the form

$$x_{n+1} = x_n + t_n d_n, \tag{4}$$

where d_n solves a linear system of the form

$$A_n d_n = -\nabla\phi(x_n) \tag{5}$$

and A_n is an $N \times N$ matrix characteristic of the method. The scalar sequence $\{t_n\}$ is usually chosen in such a way as to insure that $\phi(x_{n+1}) < \phi(x_n)$. Generally speaking, the rules for choosing $\{t_n\}$ are either ad hoc or ineffective for real problems and even the good ad hoc rules often have difficulty as $\nabla\phi(x_n)$ gets small. As a consequence, it is very desirable for the basic method to allow $t_n = 1$ for x_n close to x^*. This property is often called <u>local convergence</u>.

In fact, one of the main thrusts in current research on unconstrained minimization is to do away with the choice of $\{t_n\}$ in favor of modifications to $\{A_n\}$, [14], [16], [18], [29], [36] or by the use of various forms of continuation [1], [7], [8], [11], [15], [28]. At any rate, since the situation is more or less uniform with respect to all of the methods to be considered here, the choice of $\{t_n\}$ is somewhat beside the main points we wish to make, except as it concerns local convergence.

Another consideration, which is both computationally and theoretically interesting, is the choice of the method for solving (5), [3], [6], [20], [25], [32].

One more point to be made before we begin our treatment of specific methods is that we shall assume throughout that in some neighborhood of x^*, every $\nabla^2 f_i$ satisfies a Lipschitz condition of order one. That is, there are some constants ε, K, $K' > 0$

such that for every $i = 1,2,\ldots,M$ and any $x,x' \in N(x^*,\varepsilon) = \{x \in E^N: ||x-x^*||_2 < \varepsilon\}$,

$$||\nabla^2 f_i(x) - \nabla^2 f_i(x')||_2 \leq K||x-x'||_2$$

and

$$||J_F(x) - J_F(x')||_2 \leq K'||x-x'||_2.$$

This convention will allow us to informally state results in context without regard to specific, sometimes weaker, hypotheses used in the particular reference cited.

3. The Particular Methods.

It is fitting to begin with Newton's method.

3.1. Newton's method [31]. $A_n = \nabla^2\phi(x_n)$

ADVANTAGES:

1) It is locally convergent independent of the value of $\phi(x^*)$.

2) The convergence is ultimately second order, i.e., $||x_{n+1}-x^*|| \leq C||x_n-x^*||^2$, if $\nabla^2\phi(x^*)$ is

nonsingular.

3) The method can be used independent of the rank of $J_F(x)$.

DISADVANTAGES:

1) Every $\nabla^2\phi(x_n)$ must be nonsingular.

2) Analytic determination of J_F is required.

3) Analytic determination of $\nabla^2\phi$ is required.

In the author's opinion, the first and second advantages are the most important, since $J_F(x)$ seems to be of full rank, at least at the solution, for many practical problems.

The principal disadvantages are the second and third, since J_F is composed of NM partial derivatives. The determination of these derivatives may be undesirable for various reasons, but one certain reason is the likelihood of committing an error in determining the derivatives and programming them. Of course, these disadvantages are compounded in the computation of $\nabla^2\phi$, since J_F is used and one must either multiply out $J_F(x)^T F(x)$ and differentiate the result or determine each of the component Hessians.

These approaches are probably equally undesirable. As symbol manipulation packages become cheaper and more available, these problems will be alleviated somewhat. However, it is difficult to foresee their being completely eliminated, because the expense of evaluating the partial derivatives at each iteration would remain.

It is possible to argue that if $\phi(x^*)$ is sufficiently small, then $\sum_{i=1}^{M} f_i(x)\nabla^2 f_i(x)$ is probably not important near the solution and so the local convergence properties of the method would not suffer excessively from neglecting the term altogether. This line of reasoning is one way of arriving at the Gauss-Newton iteration.

3.2. <u>The Gauss-Newton method</u> [19].

$$A_n = J_F(x_n)^T J_F(x_n)$$

ADVANTAGES:

1) It is locally convergent if $K'\phi(x^*)^{1/2}$ is a strict lower bound for every eigenvalue of $J_F(x^*)^T J_F(x^*)$, [9].

2) The convergence rate is ultimately of order one under the condition above and of order two if, in addition $\phi(x^*) = 0$, [9].

3) No second order partial derivatives are required.

DISADVANTAGES:

1) An analytic determination of J_F is required.

2) For every n, $J_F(x_n)$ must be of full rank, but A_n may still be ill-conditioned.

3) Local convergence depends on the magnitude of $\phi(x)$ relative to the nonlinearity of F and the spectrum of $J_F(x)^T J_F(x)$, [9], [32]. More simply put, trouble can be expected when the residuals are large.

In order to obtain advantage 3) we have introduced disadvantages 2) and 3). It will be easy to overcome disadvantage 2) and we shall return to disadvantage 3) later.

Since A_n is symmetric and at least positive semidefinite (positive definite if $J_F(x_n)$ has full rank) it is easy to alter the conditioning of the system (5) by adding a scalar matrix $\mu_n I$ to A_n. This raises the problem of choosing the $\{\mu_n\}$ in such a way as to retain the local convergence properties, advantages 1) and 2), of the Gauss-Newton method. Fortunately, this isn't very difficult [9]. The following is perhaps the best and most widely used method.

3.3. <u>The Levenberg [26]-Marquardt [27] method</u>. $A_n = \mu_n I + J_F(x_n)^T J_F(x_n)$, $\mu_n \geq 0$

ADVANTAGES

1) If $\mu_n = O(||F(x_n)||)$, for any norm, then the same local convergence properties hold as for the Gauss-Newton method, i.e., advantages 1) and 2) of §3.2 are retained [9].

2) Cholesky decomposition can be used to solve the linear system (5).

3) $J_F(x_n)$ can fail to have full rank without halting the iteration.

4) The scalar μ_n can be used as a descent

parameter for ϕ and this is sometimes more convenient than using t_n, [21].

DISADVANTAGES:

1) J_F is required.

2) The method is reluctant to converge for large residual problems without a careful choice of $\{\mu_n\}$ and/or $\{t_n\}$, i.e., disadvantage 3) of the Gauss-Newton method, §3.2, remains.

The Levenberg technique is certainly not the only way of dealing with disadvantage 2) of the Gauss-Newton method [3], [6], [20], [25], [32].

Bartels, Golub, and Saunders [3] suggest a method based on an eigenvalue analysis of $J_F(x_n)$. These methods can be viewed as a practical extension of Ben-Israel's work [4], [5] and could be quite useful when μ_n is treated as a descent parameter. In [3] and [5], the constrained least squares problem is considered as another way of looking at the Levenberg-Marquardt algorithm.

No doubt some readers have not taken very seriously our concern over the need to evaluate

J_F analytically. They probably would have agreed with our concern, but would partially remove this disadvantage by using finite difference approximations to the derivatives. In fact, we recommend doing just that. However, some examples given in [17] and taken from an unpublished manuscript by P. T. Boggs and the author show that one must exercise care in choosing the step size of the finite difference or theoretical and numerical convergence to a nonminimum point may result.

Intuitively, the reasons for care in approximating J_F make sense. One is after all seeking a zero of $J_F(x)^T F(x)$. If x^* is in fact a zero of F, then any reasonable full rank approximation $\{\hat{J}_n^T\}$ to $\{J_F(x_n)^T\}$ should suffice [34]. In the more common overdetermined case, $F(x^*) \neq 0$ if $\hat{J}_n \rightarrow \hat{J} \neq J_F(x^*)$. Then we might, in fact, decrease $\phi(x_n)$ with each iteration but converge to a zero of $\hat{J}^T F(x^*)$. It seems clear that it would be useful to have a Jacobian approximation rule which is at least consistent in the sense of Ortega and Rheinboldt [31].

We now proceed to an outline of a derivative free version of the Levenberg-Marquardt iteration, which seems satisfactory from both the theoretical and practical points of view.

3.4. <u>The finite difference Levenberg-Marquardt method</u> [9]. For this method, we use the following notation. Let $h = (h^1,\ldots,h^N)$ and $\varepsilon_j \in E^N$ with $[\varepsilon_j]_i = \delta_{ij}$. Then define $\Delta F(x,h)$, an $N\times M$ matrix by

$$[\Delta F(x,h)]_{ij} = \begin{cases} \{f_i(x+h^j\varepsilon_j) - f_i(x)\}/h^j, & h^j \neq 0 \\ \partial f_i(x)/\partial x_j, & h^j = 0. \end{cases}$$

Then the system (5) is replaced by

$$(6) \qquad \{\mu_n I+\Delta F(x_n,h_n)^T \Delta F(x_n,h_n)\}d_n = -\Delta F(x_n,h_n)^T F(x_n)$$

where some rule, such as that given in [9] can be used to choose $\{h_n\}$. That rule takes into account the scaling of the problem and the distance from x_n to x^*.

ADVANTAGES:

1) No analytic partial derivatives are required.

2) If the proper strategy is used to choose $\{h_n\}$, the advantages of the Levenberg-Marquardt method of §3.3, including the convergence rate are retained. Local convergence is assured if $h_n \to 0$, as $n \to \infty$.

DISADVANTAGES:

1) Disadvantage 2) of the Levenberg-Marquardt method is retained.

2) The proper choice of $\{h_n\}$ requires some estimate of $||F(x_n)-F(x^*)||$.

In the above discussion, we used the term "proper choice of $\{h_n\}$." This is really not as vague as it sounds. The method, as noted, will be locally convergent under the same hypotheses on F as required by the Levenberg-Marquardt method, so long as $h_n \to 0$. This of course is tied to the remark made previously about the utility of a consistent approximation to $J_F(x_n)$, since a simple norm argument shows that

$||\Delta F(x_n,h_n)-J(x_n)|| = O(\|h_n\|)$. If however, one makes $\{h_n\}$ converge to zero too slowly, convergence to a minimum will be impeded. On the other hand, if $h_n \to 0$ too rapidly, cancellation errors will occur too prematurely in the computation of $\Delta F(x_n,h_n)$. Computational evidence given in [9] indicates that if one has good order of magnitude estimates for $||F(x)-F(x^*)||$, then the choice $\{h_n\} = \{O(||F(x)-F(x^*)||)\}$ will provide an algorithm whose behavior is indistinguishable from the analytic form of the Levenberg-Marquardt algorithm.

At this point, let us turn our attention to the problem of finding a method which does not require $\{\nabla^2 f_i: i = 1,2,...,M\}$, but is locally convergent without regard to the magnitude of $\phi(x^*)$. One way to obtain such a method is to try to approximate $\sum_{i=1}^{M} f_i(x_n)\nabla^2 f_i(x_n)$, since this approximation together with $J_F(x_n)^T J_F(x_n)$ would furnish an approximation to $\nabla^2\phi(x_n)$. One would then hope that the resulting Newton-like method would share with Newton's method the property of local convergence independent of the relative magnitude of $\phi(x^*)$.

The most obvious way to approximate $\sum_{i=1}^{M} f_i(x_n)\nabla^2 f_i(x_n)$ would be to make finite difference approximations to the little N×N component Hessians. However, even if one used the fact that each $\nabla^2 f_i(x_n)$ is symmetric, this would require [MN(N+1)/2] difference quotients. Next, one might try making a finite difference approximation to $\nabla^2\phi(x_n)$ directly. In other words, forget that in $J_F(x_n)^T J_F(x_n)$ one already has part of the Hessian and proceed to compute $\Delta\nabla\phi(x_n,h_n)$. This time, again using symmetry, it is only necessary to compute N(N+1)/2 difference quotients. Unfortunately, each one requires a separate evaluation of $J(x)^T F(x_n)$.

In [10], K. M. Brown and the author suggested using update methods, specifically Broyden's method, to approximate each $\nabla^2 f_i(x_n)$. For $s_n = t_n d_n$, the specific form of the approximation is

$$B_i^{n+1} = B_i^n + (u_i s_n^T/[s_n^T s_n]) \tag{7}$$

$$u_i = \nabla f_i(x_{n+1}) - \nabla f_i(x_n) - B_i^n s_n .$$

Notice that ∇f_i is the i-th row of J_F. We can now show that the resulting method, characterized by $M_n = \{\sum_{i=1}^{M} f_i(x_n)B_i^n + J_F(x_n)^T J_F(x_n)\}$ is locally Q-superlinearly convergent regardless of the value of $\phi(x^*)$, in the sense that x_o and B_i^o must be close to x^* and $\nabla^2 f_i(x_o)$. Powell [35] has given the following symmetric form of Broyden's approximation

$$P_i^{n+1} = \{P_i^n + [(u_i s_n^T + s_n u_i^T)/(s_n^T s_n)] - [(s_n^T u_i s_n^T s_n)/(s_n^T s_n)^2]\}, \tag{8}$$

which only requires the storage of the upper or lower triangular portion of P_i^n, $i = 1,2,\ldots,M$.

This method is presently undergoing further tests, but it seems worthwhile for large residual problems.

3.5. <u>The Brown-Dennis method</u> [10].

$A_n = \sum_{i=1}^{M} f_i(x_n)P_i^n + J(x_n)^T J(x_n)$, $P_i^o = \Delta\nabla f_i(x_o,h_o)$, $i = 1,2,\ldots,M$.

ADVANTAGES:

1) The method is locally convergent to x^*, regardless of the value of $\phi(x^*)$.

2) The convergence is Q-superlinear and, if $\phi(x^*) = 0$, then the convergence rate is second order.

3) The method can be used independent of the rank of $J_F(x_n)$.

DISADVANTAGES:

1) It requires storage of $\{NM(N+1)/2\}$ additional matrix elements over the other methods considered.

2) The method requires that A_n be nonsingular at every iteration.

3) The analytic determination of J_F is required.

The last disadvantage is probably temporary, since some clever finite difference strategy can most likely be developed along the lines mentioned above. The method is probably already preferable to Newton's method, since a drop from second order to superlinear convergence seems a small price to pay for dispensing with $\nabla^2\phi$. On the other hand, storage might be a real problem. However, it will probably be possible to adapt some technique, such as that given by Broyden [13] or Schubert [37], to obtain a form of the update formula for P_i which would preserve any sparseness present in the $\nabla^2 f_i(x)$. This presence often occurs in curve fitting problems and could be utilized in approximating $\nabla^2 f_i(x_n)$ by $\Delta\nabla f_i(x_n,h_n)$.

Another algorithm for the approximation of the $\sum_{i=1}^{M} f_i(x_n)\nabla^2 f_i(x_n)$ portion of $\nabla^2\phi(x_n)$ is currently being tested by Charles Broyden and the author. We do not set forth the advantages and disadvantages of the method, since it is not yet clear what they are. Hence, we content ourselves with a description of the $\nabla^2\phi(x_n)$ approximation rule.

Let the approximation rule be written in the form

$$A_n = G_n + J_F(x_n)^T J_F(x_n). \tag{9}$$

We wish to choose G_{n+1} so that the quasi-Newton

equation is satisfied by A_{n+1}, i.e., for

$$(10)\qquad \begin{aligned} s_n &= t_n d_n, \\ A_{n+1} s_n &= \nabla\phi(x_{n+1}) - \nabla\phi(x_n). \end{aligned}$$

But this is equivalent to

$$(11)\qquad G_{n+1} s_n = \{J_F(x_{n+1})^T F(x_{n+1}) - J_F(x_n)^T F(x_n) - J_F(x_{n+1})^T J_F(x_{n+1}) s_n\}.$$

It is now a simple matter to construct an update formula for G_n. For example, Broyden's update translates to

$$(12)\qquad G_{n+1} = \{G_n + [J_F(x_n)^T F(x_{n+1}) - J_F(x_n)^T F(x_n) - \{J_F(x_{n+1})^T J_F(x_{n+1}) + G_n\} s_n] s_n^T / (s_n^T s_n)\} .$$

Since $\sum_{i=1}^{M} f_i(x_n) \nabla^2 f_i(x_n)$ is symmetric, it might be desirable to do the additional arithmetic to use Powell's symmetric form of Broyden's method so that G_n will be symmetric, although storage is no problem because G_n is N×N. The savings in storage over the

Brown-Dennis method will be at least a factor of M regardless of the specific update method used. It remains to be seen if the behavior of this algorithm is sufficiently close to that of the previous one before this becomes more than an empty advantage.

4. Conclusions.

It should be clear from the above that the nonlinear least squares problem is far from being solved. The intent here has been to give one view of what needs to be done. We certainly feel that large residual problems are of prime importance. These are usually the problems which have the added complication of a large number of equations. In the linear case, such problems have become fairly routine through the use of the so-called Kalman filter. (For an algebraic exposition of this technique, see [22].) Little seems to be known about the adaptability of Kalman filtering to large nonlinear problems.

Recently there has been a flurry of activity

concerning the exploitation of any linearity in the problem. For example, Golub and Pereyra [21] give a program listing, as well as an extensive bibliography on the topic.

Finally, we remark that the ideas of Powell [35], [36] and Steen and Byrne [39] coupled with the Hessian approximation, (9)-(12), are very promising.

REFERENCES

[1] J. Avila, Continuation methods for non-linear equations, Tech. Rept. TR-142, Computer Science Center, U. of Maryland, College Park, 1971.

[2] Y. Bard, Comparison of gradient methods for the solution of nonlinear parameter estimation problems, SIAM J. Numer. Anal., 7 (1970) 157-186

[3] R. H. Bartels, G. H. Golub, and M. A. Saunders, Numerical techniques in mathematical programming, Nonlinear Programming, J. B. Rosen, O. L. Mangasarian, and K. Ritter, eds., Academic Press, N. Y., 1970, 123-176.

[4] A. Ben-Israel, A Newton-Raphson method for the solution of systems of equations, J. Math. Anal. Appl., 15 (1966) 243-252.

[5] A. Ben-Israel, On iterative methods for solving nonlinear least squares problems over convex sets, Israel J. Math., 5 (1967) 211-224.

[6] A. Bjorck and G. H. Golub, Iterative refinement of linear least squares solutions by Householder transformations, BIT, 7 (1967) 322-337.

[7] P. T. Boggs, The solution of nonlinear operator equations by A-stable integration techniques, SIAM J. Numer. Anal., 8 (1971) 767-785.

[8] W. E. Borsarge, Jr., Infinite dimensional iterative methods, Pub. 320-2347, IBM, Houston, Texas, 1968.

[9] K. M. Brown and J. E. Dennis, Jr., Derivative-free analogues of the Levenberg-Marquardt and Gauss algorithms for nonlinear least squares approximation, Numer. Math., 18 (1972) 289-297.

[10] K. M. Brown and J. E. Dennis, Jr., New computational algorithms for minimizing a sum of squares of nonlinear functions, Rept. 71-6, Computer Science Dept., Yale U., New Haven, Conn., 1971.

[11] C. G. Broyden, A new method for solving nonlinear simultaneous equations, Comput. J., 12 (1969) 95-100.

[12] C. G. Broyden, Quasi-Newton methods and their application to function minimization, Math. Comp., 21 (1967) 368-381.

[13] C. G. Broyden, The convergence of an algorithm for solving sparse nonlinear systems, Math. Comp., 25 (1971) 285-294.

[14] C. G. Broyden, The convergence of single-rank quasi-Newton methods, Math. Comp., 24 (1970) 365-382.

[15] D. F. Davidenko, On a new method of numerical solution of systems of nonlinear equations, Doklady Akad. Nauk. SSSR (N.S.), 88 (1953) 601-602.

[16] W. C. Davidon, Variance algorithm for minimization, Comput. J., 10 (1968) 406-410.

[17] J. E. Dennis, Jr., Algorithms for nonlinear problems which use discrete approximations to derivatives, Proc. ACM Natl. Conf. 1970, ACM, N. Y., 1970. Also Tech. Rept. 71-98, Dept. of Computer Science, Cornell U., Ithaca, N. Y., 1971.

[18] R. Fletcher, A new approach to variable metric algorithms, Comput. J., 13 (1970) 317-322.

[19] K. F. Gauss, Theoria motus corporum coelestiam, Werke, 7 (1809) 240-254.

[20] G. H. Golub, Matrix decompositions and statistical calculations, Statistical Computation, R. C. Milton and J. A. Nelder, eds., Academic Press, N. Y., 1969, 365-397.

[21] G. H. Golub and V. Pereyra, The differentiation of pseudoinverses and nonlinear least squares problems whose variables separate, STAN-CS-72-261, Computer Science Dept., Stanford University, Stanford, Calif., 1972.

[22] I. A. Gura, An algebraic solution of the state estimation problem, AIAA J., 7 (1969) 1242-1247.

[23] I. A. Gura and R. H. Gersten, Interpretation of n-dimensional covariance matrices, AIAA J., 9 (1971) 740-742.

[24] I. A. Gura and L. J. Henrikson, A unified approach to nonlinear estimation, J. Astron. Sci., 16 (1968) 68-78.

[25] L. S. Jennings and M. R. Osborne, Applications of orthogonal matrix transformations to the solution of systems of linear and non-linear equations, Tech. Rept. no. 37, Computing Centre, Australian National U., Canberra, 1970.

[26] K. Levenberg, A method for the solution of certain nonlinear problems in least squares, Quart. Appl. Math., 2 (1944) 164-168.

[27] D. W. Marquardt, An algorithm for least squares estimation of nonlinear parameters, J. Soc. Indust. Appl. Math., 11 (1963) 431-441.

[28] G. H. Meyer, On solving nonlinear equations with a one-parameter operator imbedding, SIAM J. Numer. Anal., 5 (1968) 739-752.

[29] B. A. Murtagh and R. W. H. Sargent, A constrained minimization method with quadratic convergence, Optimization, R. Fletcher, ed., Academic Press, 1969, 215-246.

[30] M. Z. Nashed, Generalized inverses, normal solvability, and iteration for singular operator equations, Nonlinear Functional Analysis and Applications, L. B. Rall, ed., Academic Press, N. Y., 1971, 103-309.

[31] J. M. Ortega and W. C. Rheinboldt, Iterative Solution of Nonlinear Equations in Several Variables, Academic Press, N. Y., 1970.

[32] M. R. Osborne, Some aspects of nonlinear least squares calculations, Numerical Methods for Nonlinear Optimization, F. A. Lootsma, ed., Academic Press, London, 1972, to appear.

[33] M. J. D. Powell, A FORTRAN subroutine for unconstrained minimization, requiring first derivatives of the objective function, Rept. R64-69, A.E.R.E. Harwell, Didcot, Berkshire, England, 1970.

[34] M. J. D. Powell, A method for minimizing a sum of squares of nonlinear functions without calculating derivatives, Comput. J., 1 (1965) 303-307.

[35] M. J. D. Powell, A new algorithm for unconstrained optimization, Nonlinear Programming, J. B. Rosen, O. L. Mangasarian and K. Ritter, eds., Academic Press, N. Y., 1970, 31-65

[36] M. J. D. Powell, Recent advances in unconstrained optimization, Rept., T.P. 430, A.E.R.E. Harwell, Didcot, Berkshire, England, 1970.

[37] L. K. Schubert, Modification of a quasi-Newton method for nonlinear equations with a sparse Jacobian, Math. Comp., 24 (1970) 28-30.

[38] H. W. Sorenson, Least-squares estimation from Gauss to Kalman, IEEE Spectrum, 7 (1970) 63-68.

[39] N. M. Steen and G. D. Byrne, The problem of minimizing nonlinear functionals, I. Least squares, This Volume.

THE PROBLEM OF MINIMIZING NONLINEAR FUNCTIONALS

I. LEAST SQUARES

Norman M. Steen and George D. Byrne

1. Introduction.

In this paper we consider some theoretical and practical aspects of the problem of finding a minimum point of a nonlinear functional $F: E^n \to E$, $n > 1$. It will be assumed that the functional has continuous second partial derivatives on those portions of E^n that are of interest. The problem of minimizing functionals subject to auxiliary constraints will be considered only insofar as the functionals discussed may be regarded as already reflecting the constraining conditions. Of particular interest is the case in which the functional is a weighted sum of squares which we single out for special consideration.

The basic problem is to find an n-vector θ^* such that for all vectors in some neighborhood of θ^*, $F(\theta) \geq F(\theta^*)$. In general we cannot determine such a minimizing vector by a one-step procedure and must estimate θ^* as the limit of a convergent vector sequence $\{\theta^{(m)}\}$. In practice there are two key steps to solving such problems. One is to provide an initial estimate $\theta^{(0)}$ which is close, in some sense, to the minimizing vector θ^*.

The other step is to obtain an algorithm that will construct the remainder of the sequence once an acceptable starting point is determined. In particular, the algorithm must efficiently determine the magnitude and direction to move in order to construct the next vector at each stage of the sequence. It is this latter problem to which this paper is directed. In particular, some basic concepts involved in constructing an efficient algorithm are discussed and a new algorithm based on these ideas is described.

For the reader's convenience the paper is subdivided as follows. In §2 some basic notation and definitions are introduced. These are used in a theorem which provides conditions which are sufficient to insure convergence of a fairly general type of algorithm. A discussion of the theorem is presented in order to place the overall problem in a reasonable mathematical perspective. Section 3 deals with the use of an approximating quadratic functional as an aid to selecting the direction and magnitude in which to move in order to construct each vector in the sequence which converges to the minimizing vector θ^*. In §4 the observations made regarding what an efficient algorithm might be expected to do are summarized and the theory, relevant to the development of the new algorithm, is outlined and the algorithm is presented in the form of a flow diagram. The performance of the algorithm for least squares problems is compared to other algorithms for a set of seven benchmark problems and a real life case which motivated this work in §5.

In keeping with the theme of the conference for which this paper was prepared, the details of the theoretical development will be left to the references except where they may aid in providing a better insight into the overall problem.

2. Sufficient Conditions for Convergence.

We begin by establishing notation and providing a few definitions, most of which will probably be familiar to the reader. If u and v are n-vectors, we denote the class of all vectors $z(a) = au + (1-a)v$, $0 \leq a \leq 1$ as $L(u,v)$.

DEFINITION 2.1. A set X in E^n is said to be <u>convex</u> if for all vector pairs u,v in X all vectors $z \in L(u,v)$ are also in X.

DEFINITION 2.2. A functional $F: E^n \to E$ defined on a convex set X, is said to be <u>convex</u> if for any vector pair (u,v) in X it follows that $F(au + (1-a)v) \leq aF(u) + (1-a)F(v)$, $0 \leq a \leq 1$.

An important special property of convex functionals, which are also in $C^2(X)$, is as follows. Let θ and $\theta+\delta$ be in X, then if F is in $C^2(X)$ we may write

(2.1) $$F(\theta+\delta) = F(\theta) - (\delta,\delta_g) + 1/2(\delta,H\delta)$$

where the gradient vector δ_g is defined as

(2.2) $$\delta_g = -F'(\theta) = -(\partial F(\theta)/\partial\theta_1,\ldots,\partial F(\theta)/\partial\theta_n)^T$$

and the matrix H, which is evaluated at some $z \varepsilon L(\theta,\theta+\delta)$ in (2.1), is of the form $H = [h_{jk}]$, $j,k = 1,\ldots,n$ with

(2.3) $$h(z)_{jk} = \partial^2 F(z)/\partial\theta_j\partial\theta_k; \quad j,k = 1,\ldots,n.$$

The matrix H is known to be positive semi-definite on X if and only if F is convex on X. This property is of considerable significance in this work. The following definition is not standard.

DEFINITION 2.3. A functional F is said to be bounded-convex on a convex set X if F is in $C^2(X)$, convex on X and the matrix H of (2.1) has bounded positive spectrum on X.

If the set X is compact then the notion of bounded-convexity and convexity for functionals in $C^2(X)$ coincide. It is to be noted that bounded-convexity does not mean that the functional is necessarily bounded from above. In the case $n = 1$ with $X = E$, the parabola $f(x) = x^2$ is bounded-convex on X but is not bounded from above.

With these definitions we are now in a position to consider conditions under which an iterative algorithm may be expected to converge to a minimizing vector θ^*. The theorem presented here is a modification of that given in [11, p. 23]. The purpose of the theorem is to provide the reader, who is not familiar with the mathematical basis for convergence of an algorithm, some feeling for how "close" the initial guess should be to the minimizing vector θ^* and what directions may be reasonably considered by an algorithm in moving from one vector in the sequence to the next. A more extensive coverage of this problem may be found in [1].

In the following we shall use the notation $\delta_g^{(m)}$ to denote the gradient vector of (2.2) evaluated at $\theta^{(m)}$ Other vectors corresponding to $\theta^{(m)}$ will be similarly denoted with a superscript m.

We assume that at each stage of the iterative process the algorithm may be regarded as ultimately constructing a matrix $T(\theta)$ to obtain an incremental vector δ defined as

$$(2.4) \qquad \delta = T(\theta)^{-1}\delta_g .$$

THEOREM 2.1. Let $F: E^n \to E$ be bounded-convex on a closed convex subset $X \subset E^n$. Let T be any n×n matrix defined pointwise on X with the following properties:

1) T has bounded positive spectrum on X and
2) if $\theta \in X$ then the vector $\theta+\delta$, where δ is as in (2.4), is also in X.

Then for any vector $\theta^{(0)}$ in X there exists a vector sequence $\{\theta^{(m)}\}$ and a scalar sequence $\{\gamma^{(m)}\}$ such that:

1) $\theta^{(m+1)} = \theta^{(m)} + \gamma^{(m)}\delta^{(m)}$, $m = 0,1,\ldots$ with $\gamma^{(m)} > 0$;
2) $F(\theta^{(m+1)}) \leq F(\theta^{(m)})$ with equality if and only if $\delta_g^{(m)} = 0$; and
3) the sequence $\{\theta^{(m)}\}$ converges to a vector

θ^* in X and $F(\theta^*)$ is the global minimum of F in X.

(NOTE: We do not require continuity of T on X.)

PROOF. Let $0 < v_o \leq v_1$ be the spectral bounds of T over X and let ρ be an upper bound on the spectrum of H of (2.3) over X. That is,

$$(2.5) \qquad \rho \geq \sup\{\lambda(H(\theta)) : \theta \in X\}$$

Without loss of generality we assume that $\rho > v_o$. Since $\theta^{(0)}$ is in X by hypothesis we shall assume that $\theta^{(m)}$ is in X and use induction to complete the proof.

For any vectors $\theta^{(m)}$ and $\theta^{(m)} + \delta^{(m)}$, both in X, let

$$(2.6) \quad \rho^{(m)} = \sup\{\lambda(H(Z)) : Z \in L[\theta^{(m)}, \theta^{(m)} + \delta^{(m)}]\}$$

and note that $0 < \rho^{(m)} \leq \rho$. Then for any γ, $0 \leq \gamma \leq 1$, it follows from (2.1) and (2.6) that

$$(2.7) \qquad F(\theta^{(m)} + \delta^{(m)}) - F(\theta^{(m)})$$

$$\leq \gamma[-(\delta^{(m)}, \delta_g^{(m)}) + (1/2)\rho^{(m)} ||\delta^{(m)}||^2 \gamma].$$

Let $\gamma'^{(m)}$ be defined as

$$(2.8) \qquad \gamma'^{(m)} = \frac{(\delta^{(m)}, \delta_g^{(m)})}{\rho^{(m)} ||\delta^{(m)}||^2} > 0.$$

Then for any γ, $0 < \gamma \leq \gamma'^{(m)}$, the right hand side of (2.7) is negative. Thus, let

$$(2.9) \qquad \gamma^{(m)} = \min\{\gamma'^{(m)}, 1\}$$

and define

$$(2.10) \qquad \theta^{(m+1)} = \theta^{(m)} + \gamma^{(m)}\delta^{(m)}.$$

Then we see that $\theta^{(m+1)}$ is in X and $F(\theta^{(m+1)}) \leq F(\theta^{(m)})$ with equality holding if and only if $\delta_g^{(m)} = 0$, in which case $\theta^{(m+1)} = \theta^{(m)}$ and the sequence has converged to a minimizing vector $\theta^* = \theta^{(m)}$.

Note that $1 > v_o/\rho > 0$ and $\gamma'^{(m)} > v_o/\rho$. Since F is convex on X it has a lower bound B. For any positive integer M we see from (2.7) and (2.9) that

$$F(\theta^{(0)})-B \geq \sum_{m=0}^{M} [F(\theta^{(m)}) - F(\theta^{(m+1)})]$$

$$\geq \sum_{m=0}^{M} \gamma^{(m)} [(\delta^{(m)},\delta_g^{(m)})$$

$$- (1/2)\rho^{(m)} ||\delta^{(m)}||^2 \gamma^{(m)}]$$

$$= \sum_{m=0}^{M} \gamma^{(m)} \rho^{(m)} [\gamma'^{(m)} - 1/2\gamma^{(m)}] \bullet ||\delta^{(m)}||^2 \tag{2.11}$$

$$\geq 1/2 \sum_{m=0}^{M} \gamma^{(m)^2} \rho^{(m)} ||\delta^{(m)}||^2$$

$$\geq (v_o^2/\rho) \sum_{m=0}^{M} ||\delta^{(m)}||^2$$

$$\geq \left(\frac{v_o^2}{\rho v_1}\right) \sum_{m=0}^{M} ||\delta_g^{(m)}||^2 .$$

Since the bound on the left side of (2.11) is independent of M, all infinite series on the right side converge and hence, $||\delta^{(m)}|| \to 0$ and $||\delta_g^{(m)}|| \to 0$ as $m \to \infty$. Thus $\{\theta^{(m)}\}$, converges to a vector θ^* in X and $F(\theta^*)$ is a minimum point of F in X. Finally, since F is convex on X, it is known [11, p. 40] that $F(\theta^*)$ is the global minimum of F on X.

Q.E.D.

To interpret the possible directions of the incremental vectors at each iteration that are permitted in the theorem we may define α to be the angle between the incremental vector δ and δ_g. Then we note that

$$(2.12)\quad 1 \geq \cos\alpha = (\delta,\delta_g)/[||\delta||\cdot||\delta_g||] \geq v_o/v_1 > 0.$$

Since the spectral bounds v_o and v_1 were arbitrary, except that $0 < v_o \leq v_1$, we see that the incremental vector may be oriented in any direction within 90° of δ_g. However the scalars, $\gamma^{(m)}$, which ultimately determined the step size in the theorem are dependent on the spectral bound $\rho^{(m)}$ of the matrix H which will vary with the chosen direction. Although these particular values of $\gamma^{(m)}$ are conservative in general they give an indication that the acceptable step size must be expected to vary with direction. This is clearly born out in practice if we take the matrix

$T(\theta)$ to be the identity matrix at each iteration. In this case the algorithm in Theorem 2.1 becomes the gradient or steepest descent method which is notoriously slow to converge due to the small step size that must be taken in the δ_g direction. It is of considerable interest to note however that the theorem does indicate that a finite step may be taken in the δ_g direction. This point will be considered further in §3.

In regard to the initial guess, it is natural to inquire if it is necessary that $\theta^{(0)}$ be chosen to lie in the region where F is convex. The answer to this question is no, it is not. In fact, since the person providing the initial estimate does not usually have a means for determining whether or not the functional is convex in the immediate neighborhood of a given point, the initial estimates are often chosen outside the region where F is convex. However, disastrous things can happen when the functional is not convex at the starting point. Some useful illustrations of such instances may be seen by considering the simple function

$$(2.13)\qquad f(x) = \begin{cases} [\exp(-x/\pi)][\cos^2 x], & x \geq 0 \\ \text{undefined}, & x < 0 \end{cases}$$

which is shown in Figure 2.1. The solid horizontal lines indicate the intervals in which f is convex. Since finding a minimum point of (2.13) is the same as finding a zero of f, we may attempt to use Newton's method to find the minimum point $x^* = 3\pi/2$, starting from $x^{(0)} = \pi$. The increment $\delta^{(0)} = \pi$ is then obtained and we find that $f(x^{(0)} + \delta^{(0)}) < f(x^{(0)})$. If we now set $x^{(1)} = x^{(0)} + \delta^{(0)}$ and continue in this manner we will find that at each stage of the iterative process, $\delta^{(m)} = \pi$, $f(x^{(m)}+\delta^{(m)}) < f(x^{(m)})$ and $x^{(m)} = (m+1)\pi$, $m = 0,1,\ldots$. Thus, although the initial guess was within 50% of a minimum point, Newton's method will construct a sequence which reduces the functional at each iteration but does not converge. On the other hand, if we start slightly to the left of the peak near $x = \pi$ and try to obtain

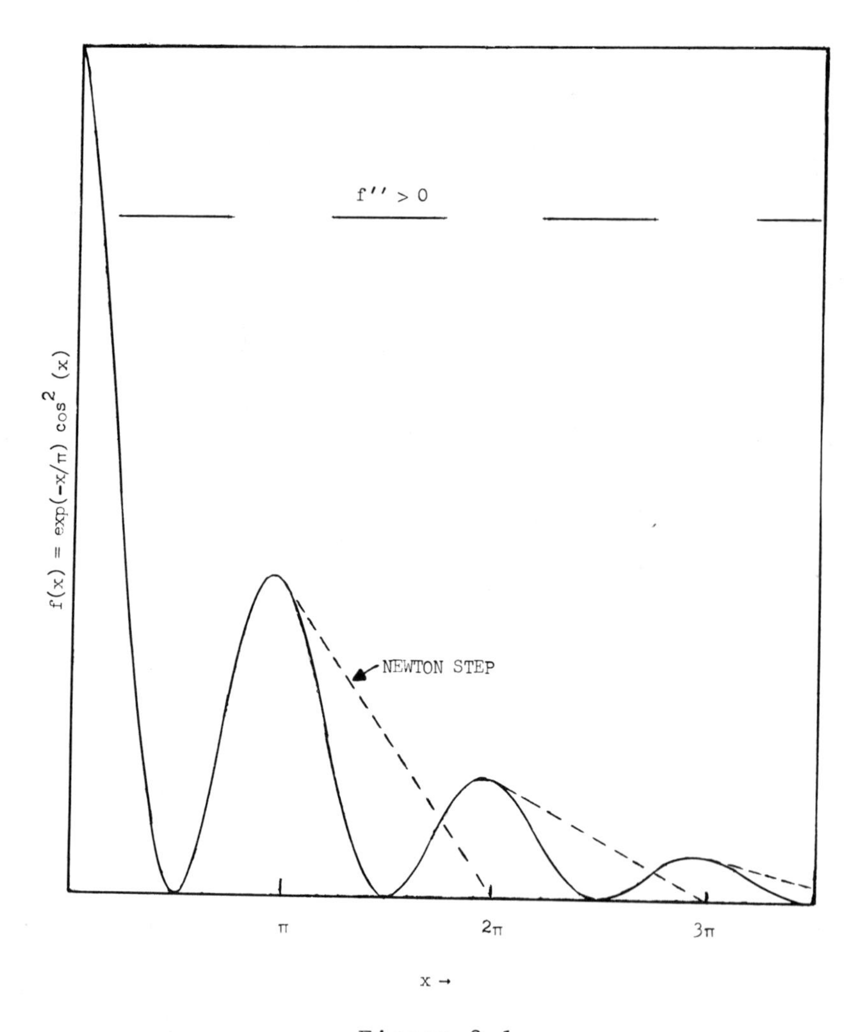

Figure 2.1

the minimum point $x^* = \pi/2$, then Newton's method will predict a point for which f is not defined. A real-life analogy of this is that the predicted point lies in a region where the functional cannot be evaluated or violates certain physical requirements of the problem. To place this example in the perspective of Theorem 2.1, the n×n matrix T simply becomes a scalar which, in the case of Newton's method, may be expressed as

$$T(x) = \delta_g^2/f(x) = \exp[-x/\pi][1/\pi \cos x + 2 \sin x]^2 \geq 0$$

where of course $\delta_g = -f'$ and equality is attained only for those values of $x = x'$ for which $\tan x' = -(2\pi)^{-1}$. However, for such values of x we may observe that

$$f''(x') = -(1/\pi^2)\exp(-x'/\pi)\cos^2(x')[1/2 + 2\pi^2] < 0.$$

Thus T is positive and bounded on each interval in which f is convex. Also, it is not hard to see that for all x in each of these intervals $x + T^{-1}\delta_g$ is also in the same interval. From this viewpoint all of the hypotheses of the theorem are satisfied except that the initial guess was not chosen in the region where the functional is convex. Consequently, even though the initial estimates were always within 50% of the solution sought, neither problem would converge. A conventional approach to handling such problems is to first try the predicted increment. If the functional is not reduced, then increments of shorter length are considered until one is found which reduces the function. However, this technique would not have helped in the first problem since the functional was reduced at each step by the incremental vector. A more reasonable approach in the authors' view is to provide an absolute constraint on the size of the increment, such as not exceeding 50% of the current estimate $x^{(m)}$ of the solution point. Such a technique will limit the step size when the functional is not convex and will permit the full increment to be used when the sequence is sufficiently close to the minimizing point x^*.

Of course we also must test to insure that the

functional has been reduced. If not, then the permissible step size must be reduced and the process repeated. This technique permits sampling the curvature of the surface more frequently when the functional is not convex. Of course there is the danger of consistently selecting too small a step which makes the algorithm inefficient. To carry this notion further we shall abandon the simple function of (2.13) and go back to the case $n > 1$.

3. The Quadratic Functional.

We noted in §2 that the direction and step size of the incremental vector must be expected to be related due to the nature of the matrix H of (2.1) and (2.3). One way in which an algorithm may attempt to interrelate these quantities is to introduce a quadratic functional which at least locally is a reasonable approximation to the true functional to be minimized. That is, for any vector θ chosen from the set X and for all incremental vectors δ, we define the quadratic functional

$$\tilde{F}(\theta+\delta) = F(\theta) - (\delta_g,\delta) + 1/2(\delta,A\delta) \tag{3.1}$$

where δ_g is as in (2.2) and the $n\times n$ matrix A is real, positive definite and evaluated at θ. If both θ and $\theta+\delta$ lie in X the difference between (3.1) and (2.1) is due to the use of the matrix $A = A(\theta)$ to approximate the matrix $H(z)$ for some $z \in L(\theta,\theta+\delta)$.

In the least squares problem the matrix A is classically obtained from the Gauss-Newton approximation which always produces a matrix which is positive semi-definite. However, other approximations of H may also be used.

Minimization of $\tilde{F}$ with respect to the elements of the incremental vector $\delta = (\delta_1,\ldots\ldots,\delta_n)$ yields the n conditions:

$$0 = \partial\tilde{F}(\theta+\delta)/\partial\delta_j = -\delta_{gj} + \sum_{k=1}^{n} a_{jk}\delta_k, \qquad j = 1,\ldots,n. \tag{3.2}$$

These conditions may be expressed in matrix notation as

(3.3) $$A\delta_t = \delta_g.$$

Since A is positive definite, the vector δ_t defined in (3.3) yields the minimum of $\tilde{F}$ for all δ and fixed θ. If $\tilde{F}(\theta+\delta) = F(\theta+\delta)$ for all δ then the vector δ_t would minimize F as well, in which case we have that $\theta^* = \theta+\delta_t$ and $X = E^n$. In general, this delightful situation does not occur and further effort is called for. In particular, referring back to Theorem 2.1, we have no assurance that $\theta+\delta_t$ is in X. Of course, we may hope to find a fraction $\gamma \varepsilon (0,1]$ such that $\theta+\delta_t' \varepsilon X$ where $\delta_t' = \gamma\delta_t$. If so, then we are assured that there is a γ' such that $F(\theta+\gamma'\delta) < F(\theta)$. This is precisely the basis of the Newton-Gauss algorithm. However, if γ' is small compared to unity, the matrix A is not locally representative of the matrix H(z). Thus it would be reasonable to expect to find a better direction in which to move.

One such approach, due to Marquardt [7] is to consider the incremental vector δ_s defined as

(3.4) $$\delta_s = (sI+A)^{-1}\delta_g, \quad s \geq 0.$$

The advantage of this choice is that of all vectors δ such that $||\delta|| = ||\delta_s||$, δ_s minimizes $\tilde{F}(\theta+\delta)$. Also, at $s = 0$, $\delta_s = \delta_t$ and as s increases $||\delta_s||$ decreases and as $s \to \infty$, $\delta_s \to (1/s)\delta_g$. In this approach, a small value of s, say s_o is tried. If the functional F is not reduced, a larger value s_1 is considered and so on until some s is found which will reduce the functional. Although Marquardt's algorithm was a major step forward there are some serious drawbacks remaining. In particular, considering our earlier remarks, we note that an increment of zero norm is predicted in the gradient direction. Also there is no convenient way to obtain control on the absolute step size (or norm) of the incremental vector. Instead, we must rely on a knowledge that as s increases the norm of the incremental vector decreases from $||\delta_t||$ at $s = 0$. Since $||\delta_t||$ varies considerably, the actual magnitude of the increment remains uncertain. An additional problem which is pointed out by Jones [6] is that for each value of s the matrix (sI+A) must be inverted, or a similar operation performed, which can be time consuming.

It is also worth noting that if H(z) is not well-represented by the matrix $A(\theta)$, then complete optimization of $\tilde{F}$ offers little if any gain over partial optimization.

Before proceeding to §4 it is worthwhile to consider the problem in which the matrix H is singular at the minimum point θ^*. When $\theta^{(m)}$ is close to θ^* the matrix $A = A(\theta^{(m)})$ should be a good approximation# to $H(\theta^{(m)})$. Thus this situation can be anticipated if the smallest eigenvalue of A is monitored and its variation between iterations is noted. If the smallest eigenvalue of A tends toward zero as $\theta^{(m)}$ approaches θ^* it may be advantageous to double the step size. When $\theta^{(m)}$ is "close" to θ^*, then the incremental vector $\delta = \delta_t$ in which case the algorithm

#This is true for least square problems when A is obtained from either the Gauss-Newton or a more accurate approximation to H.

corresponds to Newton's method for finding a zero of δ_g. If the zero is of multiplicity two then replacing δ_t by $2\delta_t$ will insure that the rate of convergence will be at least quadratic. In general, if the zero is of multiplicity q then we should consider $q\delta_t$ in place of δ_t.

4. The Algorithm.

We are now in a position to summarize what might be expected to be an efficient algorithm. Since we anticipate that the direction and step size are interrelated and also want a simple means of constructing an incremental vector it is natural to consider a linear combination of the vectors δ_t and δ_g of (3.3) and (2.1). That is, we shall consider incremental vectors of the form

$$\begin{aligned}(4.1) \qquad \delta = \gamma\delta_p &= \gamma[p\delta_g + (1-p)\delta_t] \\ &= \gamma[pI + (1-p)A^{-1}]\delta_g .\end{aligned}$$

Based on Theorem 2.1 we note that for appropriate

choices of p the vector δ is within 90° of δ_g and thus we may expect to find a γ such that $F(\theta+\delta) < F(\theta)$ provided that $\delta_g \neq 0$.

In light of our earlier remarks, the following conditions on δ would be desirable. Let r be any given fraction $0 < r \leq 1$, then:

1) $||\delta|| = r||\delta_t||$ and at $r = 1$, $\delta = \delta_t$;

2) δ is chosen in a semi-optimal manner in regard to minimizing $\tilde{F}$;

3) at $p = 1$, $||\delta|| > 0$; and

4) δ is always within 90° of δ_g.

Several recent algorithms select an incremental vector from the plane defined by δ_g and δ_t. For example, the SPIRAL algorithm [6] selects vectors from this plane but meets neither condition 2) nor 3) and no absolute control is provided on the step size. The 1970 algorithm by Powell [10] comes closest to meeting the above criteria but does not appear to satisfy condition 2).

It is possible to satisfy all of these

conditions and in a manner which is trivial computationally. Before pursuing the necessary theoretical details we make the simple observation that if both the functionals in (2.1) and (3.1) are multiplied by a scalar $c > 0$ then both the matrices H and A as well as the vector δ_g are scaled by c, but the vector δ_t remains invariant. Scaling A also scales its eigenvalues by the same factor. Hence, so long as A and H are positive definite we can always scale both (2.1) and (3.1) to insure that the smallest eigenvalue of A exceeds 2. This observation leads to the following definition.

DEFINITION 4.1. An $n \times n$ matrix is said to be of <u>type H1</u> if it is real, symmetric, and has distinct real eigenvalues which exceed 2.

Type H1 matrices play a key role in the theory behind the algorithm and are sufficient to insure that we can achieve the objectives listed above. In particular, we shall ask that the matrix A of (3.1) is of type H1. That this is not a restriction may be seen as follows. In practice A will be real and positive semi-definite hence symmetric with nonnegative eigenvalues. However, the set of positive definite matrices with distinct eigenvalues form a dense subset of the positive semi-definite matrices [11, p. 26]. Thus since we can easily scale the problem to insure that the eigenvalues exceed two we can work with the dense subset. With this definition we are now in a position to sketch the theory and the mechanism by which the scalars p and γ may be obtained. The following theorem is a condensation of several theorems which are proven in [11] and [12]. Since the proof would be rather long and not illuminating, it will be left to the references.

THEOREM 4.1. Let the matrix A of (3.1) be of type H1 and assume that $A\delta_g \neq \lambda\delta_g$. Then there exists an r_o, $0 < r_o < 1$, and a pair (γ,p) such that the following hold for any r, $r_o \leq r \leq 1$:

1) $r||\delta_t|| = ||\gamma\delta_p||$, and at $r = 1$, $(\gamma,p) = (1,0)$ (The vector $\gamma\delta_p$ is as defined in (4.1).);
2) $\partial\tilde{F}(\theta+\gamma\delta_p)/\partial p = 0$;

3) p is non-negative and increases continuously as r decreases (That is, as r decreases, $\gamma\delta_p$ moves away from δ_t.);

4) $\gamma\delta_p$ is always within 90° of both δ_g and δ_t; and

5) r_o may be chosen so that at $r = r_o$, $\delta_p = \delta_g$.

In general, the value of r_o may be taken to correspond to the limit as $p \to \infty$, in which case the direction of δ_p is $(\delta_g - \delta_t)$. For problems of the least squares type it is recommended that r_o be set so that $p = 1$ is the maximum value. The mechanism for obtaining p and γ, as well as r_o is summarized below. Define the vector σ as

(4.2) $$\sigma = \delta_g - \delta_t$$

and define the following scalars:

(4.3) $$\eta = (\sigma, A\sigma)/(\delta_g, \sigma)$$

(4.4) $$b_1 = (\delta_t, \sigma)/||\delta_t||^2$$

(4.5) $$b_2 = ||\sigma||^2/||\delta_t||^2.$$

An important consequence of the fact that A is of type H1 is the ordering $\eta > \sqrt{b_2} \geq b_1 > 1$.

In order to determine r_o we define

(4.6) $$r_1^2 = (b_2-b_1^2)/(b_2-2b_1\eta+\eta^2)$$

(4.7) $$r_2^2 = (b_2/\eta^2)$$

(4.8) $$r_3^2 = (1+2b_1+b_2)/(1+\eta)^2.$$

If we now define r_o as

(4.9) $$r_o = \max(r_1,r_2,r_3)$$

then at $r = r_o$, $\delta_p = \delta_g$.

In order to obtain p and γ for a given value of $r \geq r_o$, we proceed as follows. Define

(4.10) $$a_1(r) = b_1 - \eta r^2$$

(4.11) $$a_2(r) = b_2 - \eta^2 r^2$$

(4.12) $$V(r) = -(b_2-b_1^2) + (b_2-2b_1\eta + \eta^2)r^2$$

Then we may obtain p and γ as

$$(4.13) \qquad p = p(r) = -[a_1(r)+V(r)^{1/2}]/a_2(r)$$

and

$$(4.14) \qquad \gamma = \gamma(r) = 1/(1+\eta p).$$

If instead of the definition in (4.9) we define r_o as

$$(4.15) \qquad r_o = \max(r_1, r_2)$$

then the values of p may exceed unity and the incremental vector can take on the direction of σ as p tends to infinity. This is not generally recommended for least squares problems. Also, the value of r can be permitted to exceed unity. This is discussed in [11] and [12]. Again, however, we do not recommend this for normal least squares analysis.

Independent of whether r_o is defined by (4.9) or (4.15), if r is chosen so that $0 \le r \le r_o$ we define p and γ as follows.

$$(4.16) \qquad p = p_o = p(r_o)$$

(4.17) $\gamma = (r/r_o)/(1+\eta p_o)$.

For the reader's convenience a rough schematic of the endpoints of incremental vectors obtained with the algorithms of Marquardt [7], Jones (SPRIAL) [6] and the authors is given in Figure 4.1. This figure is for illustrative purposes only and the relative magnitudes shown may not be correct in general, except in the δ_t, δ_g, and σ directions. Also, the vectors constructed by Marquardt's algorithm do not lie in the plane defined by δ_t and δ_g. Consequently, the curve shown for Marquardt's algorithm must be regarded as a projection onto this plane.

For the least squares problem we would start at the point δ_t if $r = 1$, and as r decreases the end-point of the increment would follow the indicated curve until $r = r_o$ is encountered. For values of $r < r_o$ we then follow the δ_g direction down to the origin denoted as $\theta^{(m)}$.

There are still several important points which we must yet consider, namely, the methods to estimate

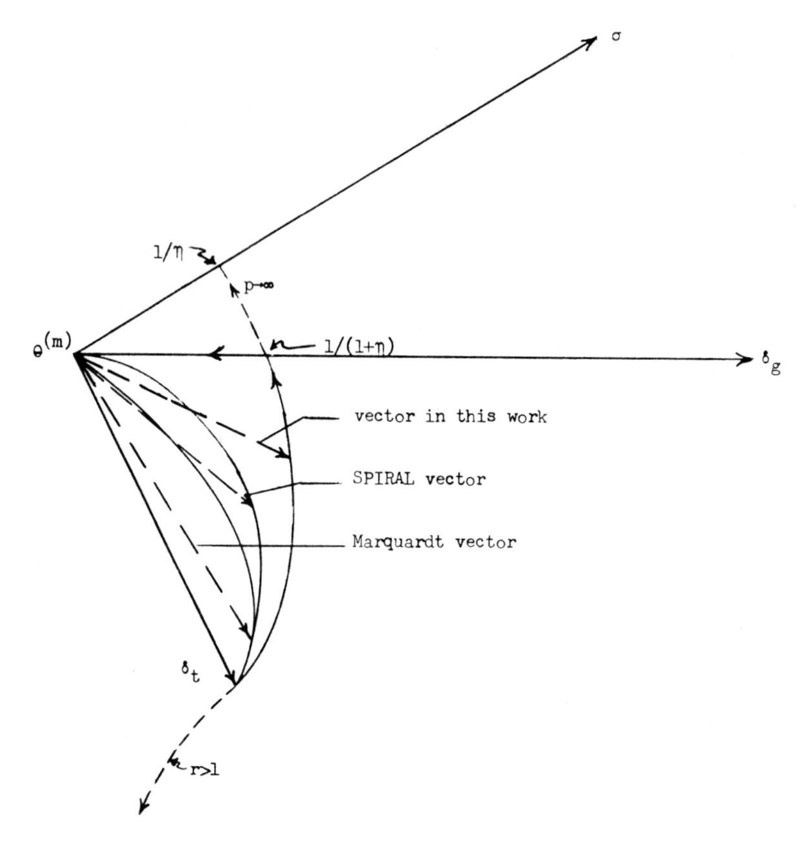
σ
1/η
p→∞
θ(m)
1/(1+η)
δg
vector in this work
SPIRAL vector
Marquardt vector
δt
r>1

Figure 4.1

the smallest eigenvalue of the matrix A and the fraction r, as well as techniques to handle the case where A is singular and when δ_g is an eigenvector of A. Each of these items is simple to carry out mechanically and we shall deal with them in the order indicated.

In regard to the smallest eigenvalue $\underline{\lambda}$ of A, all we need is a lower bound estimate $\underline{\tilde{\lambda}}$. If A is positive definite we can compute A^{-1}. If we let β be the largest eigenvalue of A^{-1} then an upper bound on β may be simply obtained as

$$\beta' = \max\{\sum_{j}^{n}|a_{jk}|: \quad 1 \leq k \leq n\} \geq \beta. \tag{4.18}$$

Then, we may define $\underline{\tilde{\lambda}}$ as

$$\underline{\tilde{\lambda}} = 1/\beta' \leq 1/\beta = \underline{\lambda}. \tag{4.19}$$

Hence by simply summing the rows (or columns) of A^{-1} we can obtain a lower bound on the smallest eigenvalue of A. This operation is trivial in terms of computational effort.

Next, we deal with the problem of estimating

the fraction r. In keeping with the spirit of our remarks in §2, we shall use a pseudo-norm $||\cdot||^*$ and a specified bound $d > 0$ on $||\delta_t||^*$ to obtain r. The reader will recall that a pseudo-norm obeys all of the properties of a norm except that $||x||^* = 0$ does not necessarily imply that $x = 0$. In principle the particular pseudo-norm that is chosen is not important. However, in practice it is of some significance. The following definitions have been successfully used

$$(4.20) \quad ||\delta_t||^* = \max\left\{\frac{|\delta_{tj}|}{|\theta_j| + \tau_j}: j = j_1, j_2, \ldots, j_k, \ k \le n\right\}$$

where the τ_j are positive constants, for instance $\tau_j = 0.1$

$$(4.21) \quad ||\delta_t||^* = \max\{|\delta_{tj}|: \ j = j_1, j_2, \ldots, j_k, \ k \le n\}$$

$$(4.22) \quad ||\delta_t||^* = ||\delta_t||_\infty .$$

To obtain r, we take

$$r = \min\{d/||\delta_t||^*, r_{max}\} \tag{4.23}$$

where for normal least squares problems $r_{max} = 1$. From (4.23) we see that when δ_t is large, the increment is kept relatively small, and as δ_t decreases we permit the full step predicted by the approximant $\tilde{F}$.

In the event a singularity of the matrix A is encountered, we can modify A by adding a scalar matrix with small norm to A to insure that the modified matrix is nonsingular.

The remaining problem, concerning the case where δ_g is an eigenvalue of A, is not really a special case. The term r_o defined in (4.9) or (4.15) is actually a measure of how far δ_g departs from an eigenvalue of A. When δ_g is an eigenvalue of A either definition of r_o yields the value $r_o = 1$ and we can proceed exactly as we have described above. When $r_o = 1$ the vectors δ_t and δ_g have the same orientation and differ only by a scale factor.

The algorithm is given in the form of a flow

diagram in Figure 4.2 which describes the algorithm for one iteration. At the outset of the problem we must provide an estimate of the parameter d which ultimately controls the magnitude of the incremental vectors. A value of $d > 0$ may be supplied, based on a knowledge or estimate of the permissible magnitude of the increments or may initially be set equal to the pseudo-norm computed in the first iteration. The diagram in Figure 4.2 includes several program constants. The requirements on these constants as well as reasonable values which have been used are summarized in Table 4.1.

TABLE 4.1

Program Constants

Constant	Requirement	Reasonable Value
ε'	$\varepsilon' > 0$	$\varepsilon' = 0.01$
c_o	$c_o > 2$	$c_o = 2.5$
c_1	$c_1 > 1$	$c_1 = 3$
ν	$\nu < 1$	$\nu = 1/2$

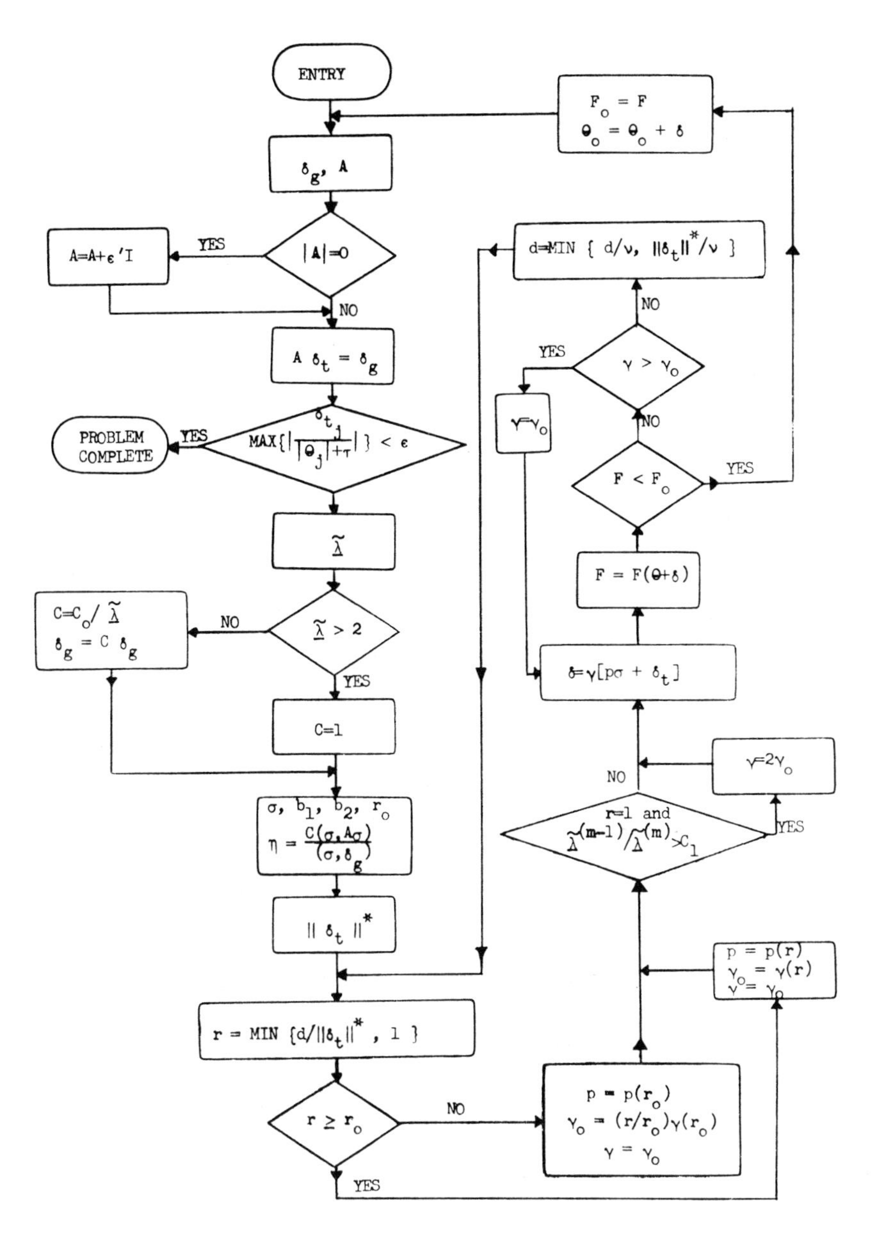
ENTRY
F_o = F
Θ_o = Θ_o + δ
δ_g, A
|A|=0
YES
A=A+ε'I
NO
A δ_t = δ_g
d=MIN { d/ν, ||δ_t||*/ν }
NO
YES
γ > γ_o
γ=γ_o
NO
PROBLEM COMPLETE
YES
MAX{| δ_tj / (|Θ_j|+τ) |} < ε
F < F_o
YES
λ̃
F = F(Θ+δ)
C=C_o/ λ̃
δ_g = C δ_g
NO
λ̃ > 2
YES
δ=γ[pσ + δ_t]
C=1
γ=2γ_o
NO
σ, b_1, b_2, r_o
η = C(σ,Aσ)/(σ,δ_g)
r=1 and
λ̃(m-1)/λ̃(m) > C_1
YES
|| δ_t ||*
p = p(r)
γ_o = γ(r)
γ = γ_o
r = MIN {d/||δ_t||*, 1 }
p = p(r_o)
γ_o = (r/r_o)γ(r_o)
γ = γ_o
r ≥ r_o
NO
YES

Figure 4.2

5. Numerical Examples.

We now consider some numerical examples for the special case in which the functional F is a weighted sum of squares. In this instance the functional is expressed as

$$(5.1) \qquad F(\theta) = \sum_{i=1}^{N} w_i[y_i - f_i(\theta)]^2, \quad N \geq n.$$

For functionals of this type the j-th element of the gradient vector δ_g of (2.2) is

$$(5.2) \qquad \delta_{gj} = -\partial F(\theta)/\partial\theta_j$$

$$= 2\sum_{i=1}^{N} w_i[y_i - f_i(\theta)][\partial f_i(\theta)/\partial\theta_j],$$

$$j = 1,\ldots,n.$$

The classic Gauss-Newton method of constructing the matrix A is to assume that for each i, the second and higher partial derivatives of f_i vanish. In this instance we approximate $f_i(\theta+\delta)$ with

$$(5.3) \qquad \tilde{f}_i(\theta+\delta) = f_i(\theta) + \sum_{k=1}^{n} [\partial f_i(\theta)/\partial\theta_k]\delta_k,$$

$$i = 1,\ldots,N.$$

If we substitute (5.3) into (5.1) evaluated at $\theta+\delta$ we obtain the approximant

(5.4) $$\tilde{F}(\theta+\delta) = \sum_{i=1}^{N} w_i \{y_i - f_i(\theta) - \sum_{k=1}^{n} [\partial f_i(\theta)/\partial\theta_k]\delta_k\}^2 .$$

Upon expanding (5.4) we see that

(5.5) $$\tilde{F}(\theta+\delta) = F(\theta) - (\delta_g,\delta) + 1/2(\delta,A\delta)$$

where δ_g is as in (5.2) and the matrix A has the elements

(5.6) $$a_{jk} = 2\sum_{i=1}^{N} w_i[\partial f_i(\theta)/\partial\theta_j][\partial f_i(\theta)/\partial\theta_k],$$

$$j,k = 1,\ldots,n.$$

Since this matrix is the product of an $n\times N$ matrix and its transpose it is always positive semi-definite and symmetric. So long as A is nonsingular then it is positive definite which is consistent with the approximant (3.1).

An alternative method of constructing an approximant is to let A be the matrix H of (2.1) and

(2.3) evaluated at $z = \theta$. In this case the elements of A are seen to be

$$(5.7) \quad a_{jk} = \partial^2 F(\theta)/\partial\theta_j \partial\theta_k$$

$$= \{-2 \sum_{i=1}^{N} w_i [y_i - f_i(\theta)][\partial^2 f_i(\theta)/\partial\theta_j \partial\theta_k]$$

$$+ 2 \sum_{i=1}^{N} w_i [\partial f_i(\theta)/\partial\theta_j][\partial f_i(\theta)/\partial\theta_k]\}$$

$$j,k = 1,\ldots,n.$$

In this instance the resulting approximant has the same form as (5.4) but the matrix A is now revised. In particular, comparing the matrix of (5.6) with that in (5.7) we see that (5.7) contains the matrix of (5.6) but also has additional terms, which arise from the fact that we have not assumed that the second partial derivatives of the f_i vanish in constructing (5.7).

In general the matrix of (5.7) results in a better approximant, however, it requires more effort to evaluate, due to the appearance of the second partial derivatives of the f_i and also has the disadvantage that it is positive semi-definite only so long as F is convex in a neighborhood of θ. Since the initial estimate of θ is often not in a region where F is convex this can cause serious problems with any of the algorithms we have mentioned in this paper, whereas the matrix of (5.6) is always at least positive semi-definite. However, it is possible to approximate (5.7) without evaluating the second partial derivatives [10] for vectors sufficiently close to θ^*. The order of convergence may be improved if such a modification is made.

We now consider some numerical examples in order to compare the computational efficiency of the algorithm described in this paper with previous algorithms for the least squares problem. The measure of efficiency used is the number of _equivalent functional evaluations_ required to converge to the solution of a given problem. This measure ignores any overhead time required by the algorithms and is based on the assumption that the overhead costs are small compared to the cost of functional evaluations.

There are two different types of problems to be considered which require different definitions of equivalent functional evaluations. In the first type, the functional is time consuming to evaluate but its first derivatives are trivial to evaluate by comparison and are simple enough that they may be written out explicitly without serious chance of error. For these problems the number of equivalent functional evaluations is simply the number of times the functional is evaluated in the flow chart of Figure 4.2, summed over all iterations. The first numerical example is of this type. The second type of problem requires that the derivatives be estimated numerically. That is, if e_j is an n-vector, then

$$\frac{\partial F(\theta)}{\partial \theta_j} \simeq \frac{F(\theta+e_j) - F(\theta)}{||e_j||}, \quad j = 1,\ldots,n$$

where e_j is zero in all but the j-th component, which is small and positive. Thus, in this case we not only need to evaluate the functional at the base and predicted points but must also evaluate F at the n points $\theta+e_j$, $j = 1,\ldots,n$. Consequently, the number of equivalent functional evaluations for this type of problem is larger by n at each iteration than in the first type.

For least squares problems in which the matrix A is obtained by the Gauss-Newton approximation in (5.6) the derivative evaluations necessary to construct the matrix are precisely those involved in computing the gradient vector $\delta_g = -F'(\theta)$, consequently no further functional evaluations are needed to construct A. This is not true in general for problems which are not of the least squares type.

We now describe the eight problems considered and then summarize the numerical results.

5.1. Analysis of fission neutron spectra.

This is a modeling problem in which a theoretical model of the energy spectrum of fission neutrons is fit to a set of experimental observations in order to identify the physical constants in the model. Because of the experimental nature of the observations, the minimum value of the functional is quite large, being on the order of 10^3.

The model consists of three sources of neutrons: those which "boil" off from each of two fully accelerated fission fragments and those which may be associated directly with the splitting (scission) process.

For each moving fragment we consider two coordinate systems. One, denoted as the lab (L) system, is stationary relative to the observer. The second system, denoted as the neutron-fragment center of mass (C.M.) system is a moving frame of reference whose origin is located at the neutron-fragment center of mass. Let E, η, E_f be respectively the neutron energy in the L and C.M. system and the kinetic energy per nucleon of the fragment. Then

$$(5.1.1) \qquad E = E_f + \eta + 2(E_f\eta)^{1/2}\mu$$

where $\mu = \cos\theta$ and θ is the angle in C.M. between the fragment velocity and neutron velocity.

We shall assume that the emission density per unit energy per unit solid angle in the C.M. system is of the form

$$(5.1.2) \qquad \psi(\eta,\mu) = \frac{\eta}{T^2} e^{-\eta/T} \frac{(1 + b\mu^2)}{2[1 + b/3]}$$

where T and b are parameters which identify the temperature and angular patterns of the emitted neutrons.

For a given fragment with energy E_f let $N_f(E)$ be the resulting emission distribution per unit energy in the (L) system. It may be shown that the distribution resulting from (5.1.2) is of the form

$$(5.1.3) \qquad N_f(E) = \frac{1}{1+b/3}\{N_o(E,E_f,T) + \frac{b}{3} N_2(E,E_f,T)\}$$

where

$$(5.1.4) \qquad N_o(E,E_f,T) = 1/2(E_fT)^{-1/2}\{S_o(X_1)-S_o(X_2)\}$$

and

(5.1.5)
$$N_2(E,E_f,T)$$
$$= \frac{3}{8}(E_fT)^{-3/2}\{S_2(E,E_f,T,x_1)-S_2(E,E_fT,x_2)\}$$

with

(5.1.6)
$$X_1 = \sqrt{E/T} + \sqrt{E_f/T}$$

(5.1.7)
$$X_2 = |\sqrt{E/T} - \sqrt{E_f/T}|$$

(5.1.8)
$$S_o(x) = 1/4\sqrt{\pi}\ \mathrm{erf}(x) - 1/2xe^{-x^2}$$

(5.1.9)
$$S_2(E,E_f,T,x)$$
$$= \{1/2\sqrt{\pi}\ \mathrm{erf}(x)[(E_f - E + 1/2T)^2 + 1/2T^2]$$
$$- xe^{-x^2}[1/2x^2T^2 + 3/4T^2 + (E_f-E)T]\}.$$

The emission distribution from the stationary source, $N_s(E)$, is assumed to be of the form

(5.1.10)
$$N_s(E) = E/T_2{}^2 e^{-E/T}s.$$

The total emission spectrum per unit energy of the model is then a function of nine physical parameters and expressed as

(5.1.11) $N(E)$

$$= \{wN_s(E,T_s)+(1-w)[rN_{f1}(E,E_{f1},T_1,b_1)$$

$$+ (1-r)N_{f_2}(E,E_{f_2},T_2,b_2)]\}$$

where w, r are the fraction of emitted neutrons arising from the stationary source and the first moving fragment respectively.

In addition to experimental knowledge of N(E) we also make use of experimental determinations of the two fragment energies E_{f_1}, E_{f_2}. Thus, we seek a nine dimensional vector,

$$\theta = (w,t_s,r,E_{f_1},T_1,b_1,E_{f_2},T_2,b_2)$$

which minimizes the functional

(5.1.12) $F(\theta)$

$$= \sum_{i=1}^{N} w_i[\phi_i-N(E_i)] + \sum_{k=1}^{2} w_{n+k}[E_{f_k} - \langle E_{f_k}\rangle]^2$$

where $\langle E_{f_k}\rangle$ is the experimental value of the kinetic energy per nucleon of the k-th fragment and $N \approx 500$.

In this problem, the pseudo-norm, $||\delta||^*$ was the form of (4.20) with $\tau = 0.1$ and $d = 1/2$.

The model discussed here is over simplified, but was found to be adequate to predict the performance of the algorithm for the model which was to be considered. Further details may be found in [3].

5.2. Rosenbrock's parabolic valley, I.

The functional to be minimized in this problem is

$$F(\theta_1,\theta_2) = 100(\theta_2-\theta_1^2)^2 + (1-\theta_1)^2 \tag{5.2.1}$$

starting with the initial guess $(\theta_1,\theta_2) = (-1.2,1.0)$. Rosenbrock [14] posed this problem because the surface contours of $(F(\theta_1,\theta_2) = \text{constant})$ follow a steep sided banana-shaped valley and (5.2.1) is difficult to minimize. The exact solution is (1.1).

5.3. Rosenbrock's parabolic valley, II.

The functional to be minimized is the same as (5.2.1). In this case the initial guess is taken as (-0.86,1,14). This problem was considered by

Jones [6] to study the effect on convergence of starting on the "inside shoulder" of the valley rather than the "outside shoulder" as in §5.2.

5.4. Zangwill's problem, I.

The functional to be minimized in this problem is

$$(5.4.1)\qquad F(\theta_1,\theta_2,\theta_3) = (\theta_1-\theta_2+\theta_3)^2 + (\theta_2+\theta_3-\theta_1)^2 + (\theta_1+\theta_2-\theta_3)^2$$

starting from $(\theta_1,\theta_2,\theta_3) = (1/2,1,1/2)$. This functional was posed by Zangwill [15].

5.5. Zangwill's problem, II.

The functional to be minimized is that in (5.4.1) but with the initial guess taken to be (100,-1,2.5) as suggested in [6].

5.6. Powell's quartic function, I.

The functional to be minimized is the four parameter system

(5.6.1) $$F(\theta_1,\theta_2,\theta_3,\theta_4)$$

$$= (\theta_1+10\theta_2)^2+5(\theta_3-\theta_4)^2+(\theta_2-2\theta_3)^4+(\theta_1-\theta_4)^4$$

starting from the point $(\theta_1,\theta_2,\theta_3,\theta_4) = (3,-1,0,1)$. This functional, due to Powell [9], exhibits difficulties near the solution point, (0,0,0,0) since the matrix A is singular at this point.

5.7. Powell's quartic function, II.

The functional to be minimized is the same as in (5.6.1) but the starting point is taken to be (3,2,1,3), [6]. It is of interest to note that the matrix A is singular at the initial guess as well as at the solution point in this problem.

5.8. Fletcher and Powell's helical valley.

In this problem the functional to be minimized is

(5.8.1) $$F(\theta_1,\theta_2,\theta_3)$$

$$= 100\{(\theta_3-10\phi(\theta_1,\theta_2))^2+(r(\theta_1,\theta_2)-1)^2\}+\theta_3^2$$

where

$$(5.8.2)\qquad 2\pi\phi(\theta_1,\theta_2) = \begin{cases} \tan^{-1}(\theta_2/\theta_1), & \theta_1 \geq 0 \\ \pi + \tan^{-1}(\theta_2/\theta_1), & \theta_1 < 0 \end{cases}$$

and

$$(5.8.3)\qquad r(\theta_1,\theta_2) = (\theta_1^2+\theta_2^2)^{1/2}.$$

The starting point in this problem is taken to be

$$(\theta_1,\theta_2,\theta_3) = (-1,0,0)$$

and the solution point is (1,0,0). This problem was suggested by Fletcher and Powell [2] and is such that the surface follows a steep sided helical valley.

5.9. Numerical results.

In Table 5.1 we give a comparison of several algorithms with the one presented here. This table indicates the number of functional evaluations which were required to obtain approximate solutions of the problems in §§5.2-5.8. The data for the Marquardt algorithm under the heading [6], as well as that for the Powell [8] and SPIRAL algorithms, were taken from [6]. The other data were compiled by the authors. The Marquardt algorithm version, which we used, employed the program constants suggested in [7]. Note the small difference in total function evaluations for the two sets of data for the Marquardt algorithm. The new algorithm provided identical results whether exact derivatives or divided differences were used.

The convergence criterion for each of these problems was taken to be

$$(5.9.1) \qquad \max \left\{ \frac{|\delta_{t,k}^{(m)}|}{|\theta_k^{(m)}|+\tau} : 1 \le k \le n \right\} < \varepsilon$$

where $\varepsilon = 0.0001$ and $\tau = 0.001$. The convergence test for Powell's algorithm is discussed by Jones [6, p. 307] and is purported to be equivalent to the criteria in (5.9.1).

For the algorithm presented here the initial choice of the parameter d and pseudo-norm were as follows. In problem 5.1 the pseudo-norm of (4.20) was used with the index taken over all nine variables and the parameter d initialized to 1/2, based on external information concerning the problem. For all other problems the pseudo-norm was taken to be the infinity norm of (4.22) with d initialized to the pseudo-norm at the first iteration. All program constants used were those given in Table 4.1.

The number of equivalent functional evaluations required to converge the eight problems are given in Table 5.1. All results cited for the other algorithms shown are based on approximating the derivatives by divided differences, except for the first problem where exact derivatives were used. The divided differences were obtained in the manner discussed by Jones. We note that Powell [10] reported 41 function evaluations to minimize the problem of §5.2.

Since the examples in Table 5.1 encompass a

rather large variety of problems it is not difficult to speculate that the algorithm has general application and can be expected to perform quite efficiently. Further improvements could be made in several ways. For example, the pseudo-norm governing the step length and the program constants of Table 4.2 might be chosen in a more sophisticated manner. An improved technique for detecting multiple zeros of δ_g would be welcome. A gain in overall efficiency in this and similar algorithms can be realized by making a judicious choice for A, the approximation to the Hessian. Both the method of choice and the choice need attention.

TABLE 5.1.

Number of equivalent functional evaluations required to solve problems.

Problem	Marquardt	[6]	Powell	SPIRAL	This Work
§5.1	51	---	---	---	8
§5.2	91	92	143	17	31
§5.3	83	72	103	27	37
§5.4	16	61	27	9	8
§5.5	16	21	52	13	8
§5.6	125	98	81	66	25
§5.7	134	103	112	76	30
§5.8	38	49	319	39	33
TOTAL §§5.2-5.8	503	496	837	247	172

REFERENCES

[1] R. M. Elkin, Convergence theorems for Gauss-Seidel and other minimization algorithms, Ph.D. thesis, University of Maryland, College Park, 1968.

[2] R. Fletcher and M.J.D. Powell, A rapidly convergent descent method for minimization, Computer J., 6 (1963) 163-168.

[3] L. Green, J. A. Mitchell and N. M. Steen, Cf-252 Prompt neutron fission spectrum, Nuclear Science and Eng., to appear.

[4] G. Hadley, Nonlinear and Dynamic Programming, Addison-Wesley, Reading, Mass., 1964.

[5] G. H. Hardy, J. E. Littlewood and G. Polya, Inequalities, Cambridge Univ. Press, London, England, 1934.

[6] A. Jones, SPIRAL - a new algorithm for nonlinear parameter estimation using lease squares, Computer J., 13 (1970) 301-308.

[7] D. W. Marquardt, An algorithm for least squares estimation of nonlinear parameters, J. Soc. Indust. Appl. Math., 11 (1963) 431-441.

[8] M.J.D. Powell, A method for minimizing a sum of squares of non-linear functions without calculating derivatives, Computer J., 7 (1965) 303-307.

[9] M.J.D. Powell, An iterative method for finding the minimum of a function of several variables without calculating derivatives, Computer J., 5 (1962) 147-151.

[10] M.J.D. Powell, A new algorithm for unconstrained optimization, Nonlinear Programming, J. B. Rosen, O. L. Mangasarian and K. Ritter, eds., Academic Press, N.Y., 1970, 31-65.

[11] N. M. Steen, An Algorithm for Nonlinear Parameter Estimation, Ph.D. thesis, University of Pittsburgh, 1972.

[12] N. M. Steen and G. D. Byrne, An algorithm for minimizing nonlinear functionals, to appear.

[13] J. Terrell, Fission neutron spectra and nuclear temperatures, Phys. Rev., 113 (1959) 527-541.

[14] H. H. Rosenbrock, An automatic method for finding the greatest or least value of a function, Computer J., 3 (1960) 175-184.

[15] W. I. Zangwill, Minimizing a function without calculating derivatives, Computer J., 10 (1967) 293-296.

QUASI-NEWTON, OR MODIFICATION METHODS*

C. G. Broyden

1. Introduction.

Let $f: D \subset R^n \to R^m$ be continuous and differentiable, and let $\{x_k\}$ be an arbitrary sequence in R^n. If, in order to simplify notation, we write f_k for $f(x_k)$, and define s_k and y_k by

(1) $$s_k = x_{k+1} - x_k$$

and

(2) $$y_k = f_{k+1} - f_k$$

then if some m by n matrix A_{k+1} satisfies

(3) $$A_{k+1}s_k = y_k,$$

---

*This work has done while the author was Visiting Research Professor, Department of Computer Science, Cornell University, and was supported by NSF Grant #GJ-27528.

it may be regarded [34] as a first divided difference of f on D. Note that, at this stage, there is no obligation upon m to be equal to n. It could be greater, as would be the case for an overdetermined system of equations, or less, if f represents the set of constraints in a constrained optimization problem. This freedom does not affect the essential idea underlying the modification methods but only the method of generation of the sequence $\{x_k\}$.

We first note that A_{k+1} is not defined by equation (3) since, if $p \in R^m$, $q \in R^n$ and $q^T s_k = 0$,

$$(A_{k+1} + pq^T)s_k = y_k.$$

Since q may lie anywhere in the (n-1)-dimension subspace orthogonal to s_k, any matrix that satifies equation (3) alone experiences a considerable amount of latitude. This may be curtailed by requiring A_{k+1} to satisfy

$$A_{k+1}s_j = y_j \tag{4}$$

for r pairs (s_j, y_j), where $r \leq n$. If $r = n$ then A_{k+1} is defined exactly provided that the n vectors s_j are linearly independent. If j takes the values k, k-1, k-2, ..., k-n+1, then we have a generalization of the sequential secant method [34]. Indeed there is no reason why r should not be greater than n, so that A_{k+1} is the solution of an overdetermined system. A_{k+1} would then be obtained by solving this system in some approximate sense, although the cost of this would probably exceed that of the updating methods we now describe.

Suppose now that we have some approximation B_k to the Jacobian of f evaluated at x_k. From x_{k+1} we are able to obtain a new (s_j, y_j) pair, namely (s_k, y_k), and it would then appear reasonable to require that the next approximation to the Jacobian, namely B_{k+1}, should satisfy equation (3). Since, if B_k is itself a good approximation to the Jacobian we would expect B_{k+1} to differ only slightly from B_k, we might expect to obtain B_{k+1} from B_k by adding a correction term. This approach has been discussed by Professor Rheinboldt in his lectures, [34]. We show now how imposing certain requirements upon the correction forces it to be rank-1. We observe that when we obtain y_k and s_k and require

$$B_{k+1}s_k = y_k \tag{5}$$

to be satisfied we are using the fact that y_k and s_k define a new approximation to the directional derivative of f in the direction s_k. In fact the new information about the Jacobian only refers to that specific direction, and no further information is available about the directional derivatives in any direction other than s_k. In particular no additional information is available about the directional derivatives in any direction orthogonal to s_k, so that it would appear reasonable to require that the approximate directional derivatives predicted by B_{k+1} in these directions should be precisely those predicted by B_k. This implies

$$B_{k+1}q = B_k q, \ \{q | q \neq 0, \ q^T s_k = 0\} \tag{6}$$

and this equation combined with equation (5) yields the unique updating formula

$$B_{k+1} = B_k - (Bs_k - y_k) \frac{s_k^T}{s_k^T s_k} . \tag{7}$$

This formula was proposed in [7] for square B_k but the above justification applies equally to the rectangular case. If f now denotes the linear function $Ax - b$, where A, the Jacobian of f, is a constant $m \times n$ matrix we may define a matrix error E_k by

$$E_k = B_k - A. \tag{8}$$

It is trivial in this case to see, since

$$y_k = As_k, \tag{9}$$

that if B_{k+1} is obtained from B_k using equation (7) then

$$E_{k+1} = E_k \left(I - \frac{s_k s_k^T}{s_k^T s_k} \right), \tag{10}$$

so that $||E_{k+1}|| \leq ||E_k||$, where $||\cdot||$ denotes the spectral norm. In this sense the new approximation to A given by equation (7) is no worse than the previous approximation and could conceivably be better.

2. Motivation and Implementation.

The motivation underlying the use of these modification methods is straightforward. Of the many algorithms in existence for solving a large variety of problems, most of the successful ones require the calculation of a sequence $\{x_k\}$ together with the associated sequences $\{f_k\}$ and $\{J_k\}$, where J_k is the Jacobian of f evaluated at x_k. Examples of these are Newton's method for nonlinear equations, the Gauss-Newton method for least-squares problems and the various methods inspired by the work of Levenberg and Marquardt. In all of these methods it is frequently quite costly to compute J_k, even if this is possible without resorting to the use of differences. If one uses differences then the work involved in obtaining an approximation to J_k is m times greater than that in evaluating f_k. Now since all these methods require the calculation of f_k the use of a modification technique enables an approximation to J_k to be computed <u>without further evaluation of f</u>. It is now quite obvious that this approximation B_k can be used in place of J_k whenever the latter occurs. It is of course unlikely that this device will reduce the number of iterations required to solve any given problem, and in fact the reverse is much more likely. It is possible though that since the work (measured in equivalent evaluations of f) per iteration is substantially reduced, the use of a modification method might result in an overall reduction in computing cost, and this possibility has been widely realized in practice.

The most obvious way of implementing a quasi-Newton method is merely to replace J_k whenever it occurs by B_k. This could be dangerous (see below) or inefficient. A more sophisticated implementation is possible when B_k is square and non-singular, as would hopefully be the case when solving nonlinear simultaneous equations. Here, instead of storing and updating B_k, the approximation to the Jacobian, one would store and update $H_k = B_k^{-1}$. The quasi-Newton analogue of Newton's method in this case would then become

$$x_{k+1} = x_k - H_k f_k \alpha_k, \tag{11}$$

where α_k is either equal to unity, or is chosen iteratively to ensure that $||f_{k+1}|| \leq ||f_k||$. The update for H_k corresponding to equation (7) then becomes

$$H_{k+1} = H_k - (H_k y_k - s_k) \frac{s_k^T H_k}{s_k^T H_k y_k} . \tag{12}$$

We now consider the performance of the quasi-Newton methods in more detail, and examine their application to particular types of problems.

3. General Nonlinear Systems.

Only two basic quasi-Newton methods have been seriously proposed to solve general systems of nonlinear simultaneous equations, and these are the secant method [3], [39] and Broyden's method [7]. In the method Rheinboldt [34] refers to as the "Sequential Secant Method", the updating formula for H is

$$H_{k+1} = H_k - (H_k y_k - s_k) \frac{q_k^T}{q_k^T y_k} \tag{13}$$

where q_k is chosen to satisfy

$$q_k^T y_j = 0, \quad k-n+1 \leq j \leq k-1.$$

This particular version of the secant method has attracted much theoretical attention [4], [5], [38] although this has probably been motivated more by the clean structure of the algorithm than by its observed performance on non-trivial problems. It is simple to show that the algorithm possesses a termination property, that is it will solve a set of n linear equations in at most n+1 steps, and Brent has shown in general that, for $n < 10$, the method is the most efficient of the set of algorithms that he considers in [5] provided that a certain determinant remains bounded away from zero. In practice though, for all but the most trivial problems, no such bound on the determinant exists and the secant method is notoriously unreliable. Attempts have been made [3] to overcome this, with some success.

Broyden's method ([7], and equations (11) and (12)), unlike the secant method, possesses no termination property and thus can at most give an approximate solution to a linear system. It is known [10] that it converges R-superlinearly (see [34] for a discussion of R-superlinear convergence) to the solution of this linear system and this result has since been extended by Broyden, Dennis and Moré [14], who have shown that if $\alpha_k = 1$ the method possesses Q-superlinear local convergence properties for general functions under conditions little stronger than differentiability combined with a non-singular Jacobian at the solution. Away from the solution the update seems to be robust with respect to choice of α_k (see equation (11)) although in theory a division by zero while updating is possible. If, on the other hand, B_k approximates sufficiently closely to J_k, this cannot occur and the method is numerically stable. Indeed one would expect that locally the method would be robust with respect to rounding error. If $f \equiv Ax - b$ and R_k is defined by

$$R_k = H_k A - I \tag{14}$$

then, if $||R_k|| < 1$ and H_{k+1} is given by

$$H_{k+1} = H_k + C_k$$

it may be shown [10] that

$$\frac{||C_k||}{||H_k||} \leq \frac{||R_k||}{1 - ||R_k||} .$$

Thus if $||R_k|| << 1$ the relative correction to H_k is itself small and errors in the calculation of C_k

would have comparatively little effect upon H_{k+1}. It would thus appear that the theory of Bartels (quoted by Professor Rheinboldt in his lectures [34]) might not be applicable in this case, and this view is strongly supported by the experimental evidence to date.

One feature of this method is that the choice of α_k to ensure that $||f_{k+1}|| \leq ||f_k||$ has been known to inhibit convergence near the solution. This is discussed more fully in [10], [12], and a possible way of overcoming this problem has been suggested in [9].

It might, at this point, be appropriate to mention Broyden's second method [7]. The update is given by equation (13) with $q_k = y_k$, and if α_k is chosen to reduce $||f_k||$ the method appears to be virtually useless. Curiously enough, however, if $\alpha_k = 1$, the method is Q-superlinearly locally convergent [14]. We mention this fact in support of our contention that robustness away from the solution is as important in practice as rate of convergence close to it.

If our nonlinear system is sparse, i.e. if many of the elements of the Jacobian are known to be always zero, then, as has been pointed out in the main lectures [34], in order to capitalize upon the structure it is necessary to work with an approximation to the Jacobian and not to its inverse. However, all of the general updates, being themselves non-sparse, are unsuitable as they stand and it is necessary in order to maintain a sparse approximation to J_k, to construct a sparse correction. One way of doing this was suggested by Schubert [36]. If B_k is the sparse approximation to J_k, and s_k and y_k have been computed, Schubert's procedure is as follows.

(1) Construct the Broyden correction C_k, where

$$C_k = (y_k - B_k s_k) \frac{s_k^T}{s_k^T s_k} .$$

(2) This correction is non-sparse, so

replace all the elements of C_k corresponding to known zero elements of J_k by zero. Denote this sparse correction by C_k^1.

(3) Since, in general, $B_k + C_k^1$ will not satisfy the quasi-Newton equation, i.e.

$$(B_k + C_k^1)s_k \neq y_k,$$

scale each row of C_k^1 individually, i.e. determine the diagonal matrix D_k, so that

$$(B_k + D_k C_k^1)s_k = y_k.$$

Then

$$B_{k+1} = B_k + D_k C_k^1.$$

Locally this method appears to have the same kind of convergence properties as the parent method. Convergence is at least linear [13] and we conjecture that it is superlinear.

The method has solved test problems of up to

600 equations, but as it is comparatively new (1970), few reports of its performance on actual problems have yet become available.

4. Least Squares Problems.

Of the two methods that we describe here, the first is applicable to the overdetermined system ($m > n$) but the second applies only to the case where $m = n$. We deal with the latter here rather than in the previous section as it views the problem of solving the system $f = 0$ from the standpoint of zeroing $f^T f$.

If we define ϕ by

$$\phi = (1/2) f^T f \tag{15}$$

we see that, if $g = \text{grad}\ \phi$,

$$g = J^T f \tag{16}$$

where J is the Jacobian of f. Thus finding a stationary point of ϕ is equivalent to finding a value of x for which g = 0, i.e.

$$J^T f = 0. \tag{17}$$

Thus we are faced with the problem of solving equation (17). Note that if m = n (J is square) and J is also non-singular, (17) reduces to f = 0, and we have straightforward nonlinear equations. To solve equation (17) by Newton's method requires the Jacobian of g, which we denote by K. Differentiation of (16) then gives

$$K = J^T J + \sum_{i=1}^{m} f_i G_i, \tag{18}$$

where G_i is the Hessian of f_i, the ith component of f. Since the matrices G_i are difficult to obtain, the second term on the right-hand side of (18) is usually omitted, and if J_k is replaced by B_k in the customary manner we obtain the quasi-Newton analogue of the Gauss-Newton method, namely

$$x_{k+1} = x_k - (B_k^T B_k)^{-1} B_k^T f_k \alpha_k. \tag{19}$$

For more details of the Gauss-Newton method see [26] or [34]. For more discussion of quasi-Newton analogues see [26].

Powell's method [29] for minimizing the sum of squares without using derivatives uses essentially equation (19), updating the B_k's with a modified secant update (see Section 3, above). The method has enjoyed some success and the only competing methods to date (in the sense of using the same information about the function, i.e. evaluations of f alone) are those using a finite-difference approximation to J_k. These have been shown [6] to be quadratically convergent if the minimum sum of squares is zero but it is possible that the m extra evaluations of f required at each stage in order to approximate J_k might make them slower in practice for some problems.

The principal, and dangerous, disadvantage of using equation (19) lies in the determination of a convergence criterion. The solution of $J^T f = 0$ is not in general the solution of $B_k^T f = 0$, where B_k only approximates to J, and therein lies one of the difficulties of solving the overdetermined least-squares problem. If J_k is not available, in order to test for convergence a sufficiently accurate approximation to J_k must be used, and it is by no means clear that the approximations generated by an updating procedure are sufficiently accurate. For this reason it would appear that finite-difference approximations to J_k are currently the safest ones to use.

If m = n, and the Jacobian at the solution is non-singular, then the convergence criterion may be based upon $||f||$. In this case since any errors in the approximation to J_k can only affect the speed of convergence, and not the decision to terminate the iteration, the worst that can happen is a known failure. We could then use Powell's hybrid method [30], which combines the use of an updating formula with techniques reminiscent of the Levenberg-

Marquardt method (see [26] or [34]). The method stores and updates approximations to both J_k and J_k^{-1}, using the former to generate an approximate steepest-descent direction of $f^T f$ and the latter to obtain an approximation to the Newton direction $J_k^{-1} f_k$. These are then combined to give a step which reduces $||f||$ and does not require the solution of linear equations. For further details we refer the reader to the original paper [30].

5. The Unconstrained Optimization Problem.

Of all the problems tackled by quasi-Newton methods the unconstrained optimization problem has received the most theoretical attention, has been subject to the most extensive experimentation and can furnish the greatest number of examples of genuine problems (as opposed to constructed test problems) successfully solved. It is impossible here to do more than outline the current state of the art, and the fact that we are able to attempt even this is due in large measure to the recent consolidation of the subject at the hands of Huang, Dixon and others.

The problem then is the minimization of $\phi(x)$, where f, the gradient of ϕ, is available as an explicit expression. J, the Jacobian of f, is the Hessian of ϕ and subject to the usual caveat on continuity is thus symmetric. If ϕ is expanded about an arbitrary x_0 as a Taylor series it may be seen that, for x sufficiently close to x_0, ϕ may be approximated by a quadratic function. For this reason most of the theoretical work to date on the performance of optimization algorithms has been concerned with their performance on the model quadratic problem, where

$$\phi_q = (1/2)x^T Ax - b^T x + c. \tag{20}$$

If A is symmetric and positive definite, ϕ_q possesses a unique minimum and unless an algorithm can determine this efficiently there is little likelihood that it will perform well on the more general problem.

Now certain algorithms are able to minimize ϕ_q exactly in at most n steps, and these are said to possess the quadratic termination property. Since a termination property certainly guarantees efficiency when solving quadratic problems it has often been regarded as a desirable property for an algorithm to possess, and consequently much work has been carried out on such algorithms. One particular very broad class of algorithms that possesses this property is the Huang class, which we now discuss in some detail.

Huang [23] considered the (at most rank 2) updates of the approximation H_k to the inverse Jacobian given by the formula

(21a) $$H_{k+1} = H_k + \rho_k s_k p_k^T - H_k y_k q_k^T$$

where

(21b) $$p_k = s_k \alpha_k + H_k^T y_k \beta_k$$

and

(21c) $$q_k = s_k \gamma_k + H_k^T y_k \delta_k .$$

The scalar ρ_k is arbitrary and the constants α_k,

β_k, γ_k and δ_k are arbitrary but subject to the restriction that

(21d) $$p_k^T y_k = q_k^T y_k = 1.$$

We note that

(a) Although H_k approximates to a symmetric matrix it is not required to be symmetric itself.

(b) The equation

(22) $$H_{k+1} y_k = s_k \rho_k$$

holds for all k.

(c) There are effectively three arbitrary parameters, although this is reduced to two if we require the quasi-Newton equation to hold (i.e. $\rho_k = 1$).

Using this form of the update, we compute x_{k+1} by

(23) $$x_{k+1} = x_k - H_k^T f_k \alpha_k$$

where α_k is chosen to either minimize or reduce the function ϕ along s_k. Note that if H_k is symmetric equations (11) and (23) are identical.

The Huang class contains all the popular minimization updates, and most of the little-used ones as well. One might mention the Davidon-Fletcher-Powell [16], [21], the single-rank symmetric [17], [27], Greenstadt's first [22], discussed by Professor Rheinboldt [34], and the single-rank methods of Pearson and McCormick [28]. The importance of the class lies in the following remarkable properties shared by all its members.

(a) If the iterations are perfect, i.e. α_k is chosen to minimize ϕ along s_k, then for the quadratic function defined in (20)

(a1) The algorithms terminate giving the exact solution in at most n steps

(a2) $H_k y_j = s_j \rho_j, \qquad j < k$

(a3) $s_k^T A s_j = 0, \qquad j \neq k.$

Moreover, if the process requires n steps to terminate, and the final matrix is denoted by H_n, then

(a4) $H_n = A^{-1}$ if $\rho_j = 1, \quad \forall_j,$

$H_n = 0 \quad$ if $\rho_j = 0, \quad \forall_j.$

Thus if H_k satisfies the quasi-Newton equation its final value is the exact inverse Hessian of ϕ_q. These results were proved by Huang [23].

(b) If the iterations are perfect then all algorithms sharing the same ρ_k at each step generate, for continuous ϕ, the identical sequence of iterates $\{x_k\}$ provided that some very weak restrictions are satisfied. (Dixon [19].)

The implications of Dixon's theorem are wide-reaching. The first implication is that Powell's convergence results ([32], [23], [34]) apply to all members of the Huang class provided that each iteration is perfect. Secondly, in many reported experiments widely different sequences $\{x_k\}$ have been obtained for the same problem using different algorithms, hence either the iterations have been less than perfect or the whole process is highly susceptible to rounding error. This last ambiguity has not yet been resolved. The third implication concerns the infeasibility of performing a perfect iteration. That such is possible is shown by the work of Huang and Levy [24] who obtained identical sequences $\{x_k\}$ for four different algorithms, but who performed extremely accurate line searches and worked in double-precision in order to do so. Since both these procedures are expensive it is likely that the emphasis will shift to imperfect methods, making much of the termination theory, which relies on perfect line-searches, of academic interest.

Another casualty of the departure from exact line searches is the confidence placed in the stability of certain methods. Virtually all the popular

Huang updates are also members of that subclass obtained by putting $\rho_k = 1$ in equation (21) and choosing α_k, etc., to preserve symmetry. This gives the single-parameter family introduced by Broyden [8] for which the update may be written

$$(24a) \qquad H_{k+1} = H_k - \frac{H_k y_k y_k^T H_k}{y_k^T H_k y_k} + \frac{s_k s_k^T}{s_k^T y_k} + \rho_k v_k v_k^T$$

where

$$(24b) \qquad v_k = H_k y_k - s_k \left(\frac{y_k^T H_k y_k}{s_k^T y_k} \right)$$

and ρ_k is arbitrary.

Putting $\rho_k = 0$ gives the Davidon-Fletcher-Powell method and choosing $\rho_k = (y_k^T H_k y_k)^{-1}$ gives an update recommended independently by Broyden [11], Fletcher [20] and Shanno [35].

Fletcher and Powell [21] showed that when their method is applied to a general function and an exact line search is carried out, then H_{k+1} will be positive definite when H_k is positive definite. A similar result was proved for the class (24) for $\rho_k \geq 0$ by Broyden [8] and this was extended by Shanno [35]. This retention of the property of positive definiteness, however, was observed in practice to break down, an occurrence attributed by some, including the author, to excessive rounding error, and to others, e.g. Bard [2], to poor scaling. In practice this breakdown would be circumvented by periodic "resetting" of H_k. It now seems likely that at least some of the numerical instabilities associated with the Davidon-Fletcher-Powell and similar algorithms may be attributed to inexact line searches [1], and new strategies are now being devised [20] to take this into account.

The best algorithms of the Huang class have been and are being used extensively to solve genuine problems, making nonsense of any suggestion that more theoretical work is needed before we should consider actually using them. Even if the exact line search requirement is replaced by something less stringent, the algorithms still perform well, and it has recently been proved [14] that if $\alpha_k = 1$ both the Davidon-Fletcher-Powell and the Broyden-Fletcher-Shanno algorithms are locally superlinearly convergent. Since recent tendencies [20] have been to take $\alpha_k = 1$, if this results in a reduction of ϕ, the theoretical results quoted above are in line with present computing practice as well as being of some intrinsic interest.

6. Symmetrization.

The assumption that if J is known to be symmetric then it is desirable to use a symmetric approximation to it has profoundly influenced the development of optimization methods. The early work of Davidon [16], Fletcher and Powell [21] and Broyden [8] acknowledged this assumption and it was not until Pearson's work in 1969 that the possibility of non-symmetric approximations to the Hessian was considered.

More recently Powell [31] observed that Broyden's single-rank update [7] could be symmetrized in the following way. Let B_k be a symmetric approximation to the Hessian of ϕ, and let $B_{k+1}^{(1/2)}$ be given by

$$B_{k+1}^{(1/2)} = B_{k+1} - \frac{v_k s_k^T}{s_k^T s_k} \tag{25a}$$

where

$$v_k = B_k s_k - y_k . \tag{25b}$$

Since $B_{k+1}^{(1/2)}$ is not symmetric Powell defined $B_{k+1}^{(1)}$ by

$$B_{k+1}^{(1)} = (1/2)\left(B_{k+1}^{(1/2)T} + B_{k+1}^{(1/2)}\right). \tag{26}$$

Now $B_{k+1}^{(1)}$ does not, in general, satisfy the quasi-Newton equation, <u>i.e.</u>

$$B_{k+1}^{(1)} s_k \neq y_k$$

so it too may be updated by

$$B_{k+1}^{(1+1/2)} = B_{k+1}^{(1)} - \frac{v_k^{(1)} s_k^T}{s_k^T s_k},$$

where

$$v^{(1)} = B_{k+1}^{(1)} s_k - y_k,$$

and $B_{k+1}^{(2)}$ is obtained from $B_{k+1}^{(1+1/2)}$ using an analogue of equation (26). Powell showed that this procedure carried out indefinitely gives the limiting value of $B_{k+1}^{(i)}$ to be

$$\lim_{i\to\infty} B_{k+1}^{(i)} = B_k - \frac{1}{s_k^T s_k}\left(v_k s_k^T + s_k v_k^T - \frac{s_k (v_k^T s_k) s_k^T}{s_k^T s_k} \right), \tag{27}$$

a rank-2 symmetric update which is <u>not</u> a member of the Huang class. An interesting property of this update emerges when it is applied to the function $f = Ax - b$, where A is a constant symmetric matrix. The error equation (10) is replaced by

$$E_{k+1} = \left(I - \frac{s_k s_k^T}{s_k^T s_k} \right) E_k \left(I - \frac{s_k s_k^T}{s_k^T s_k} \right) \tag{28}$$

so again we have $||E_{k+1}|| \leq ||E_k||$.

If we now symmetrize the general single-rank update

$$(29) \qquad B_{k+1} = B_k - v_k q_k^T.$$

where q_k is arbitrary save that $q_k^T s_k = 1$, we obtain equation (27) but with some of the s_k's replaced by q_k. Updates obtained by this procedure for particular choices of q_k have been discussed by Dennis [18], and one was found to be identical to one introduced by Greenstadt [22], and obtained using variational considerations. This update too is not a member of the Huang class. As was pointed out in Professor Rheinboldt's lectures [34], the operations of symmetrization and inversion do not commute, i.e. if we obtain the inverse update corresponding to equation (29) and symmetrize this by a procedure similar to Powell's we do not, in general, obtain the same update as would be obtained by symmetrizing update (29) and obtaining the corresponding inverse update.

It follows from this that the two Broyden single rank updates (Section 3, above) give rise to four double-rank formulae, two being the ones of Powell and Greenstadt mentioned above. These are known to be locally superlinearly convergent if $\alpha_k = 1$, [14], but for the remaining two, obtained by Dennis, this property remains for the moment conjectural.

The only single-rank update whose symmetrization has not yet been investigated is the secant method, but it can only be a matter of time before this is accomplished.

Whether or not the idea of symmetrization will result in the sort of consolidation of the subject achieved by Huang and Dixon is at the moment an open question. It is not yet established that every symmetric double-rank update may be obtained from some single-rank one by a Powell symmetrization, and it is possible that two single-rank updates may give the same double-rank one. In any case the addition of the term

$$\lambda_k (v_k s_k^T s_k - s_k v_k^T s_k)(s_k^T s_k v_k^T - v_k^T s_k s_k^T),$$

where λ_k is arbitrary, to the right-hand side of equation (17) affects neither the symmetry of B_{k+1} nor its ability to satisfy the quasi-Newton equation, and a comparable modification is possible in the general case. If this were done every single-rank algorithm would give rise to a class of symmetric double-rank ones, particular algorithms of the class being specified by the choice of λ_k.

7. Other Applications.

The principle of using some sequence $\{x_k\}$ and the associated sequence $\{f_k\}$ to generate an approximate Jacobian can be extended to generating approximate partitions of Jacobians, or components of Jacobians. Two examples of the former were given by Broyden and Hart [15], and Kwakernaak and Strijbos [25], in connection with the constrained minimization problem. The solution of the problem of minimizing $\phi(x)$ subject to the constraints $c(x) = 0$, $x \in R^n$,

$c \varepsilon R^m$, $m < n$, may be expressed as the solution of the set of $m + n$ nonlinear equations

$$J^T(x)z = f(x) \tag{30a}$$

$$c(x) = 0 \tag{30b}$$

where $f(x)$ is the gradient of $\phi(x)$, $J(x)$ is the Jacobian of $c(x)$ and z is the vector of Lagrangian multipliers. The Jacobian J_s for this system has the form

$$J_s = \begin{bmatrix} G & J^T(x) \\ J(x) & 0 \end{bmatrix} \tag{31}$$

where G is a linear combination of the Hessians of $\phi(x)$ and of each individual constraint function. Broyden and Hart used approximations to both G and $J(x)$ in approximating J_s whereas Kwakernaak and Strijbos approximated G alone. For a fuller description of these algorithms see [26], or the original papers.

A second application due to Hart and Soul [37] arises from the solution of ordinary differential equations using collocation methods. In the method proposed for solving the resulting nonlinear equations the Jacobian J had the form

$$J = M_0 + D_1 M_1 + D_2 M_2 \tag{32}$$

where M_0, M_1 and M_2 are known n×n matrices and D_1 and D_2 are unknown diagonal matrices. These latter were approximated by diagonal matrices, and the approximations were updated so that the new Jacobian approximation satisfied the quasi-Newton equation.

In all the above cases, [15], [25], and [37], good local convergence properties were observed, although the algorithms developed might not be optimal for the solution of the particular problems quoted. They do illustrate, however, one way in which special-purpose quasi-Newton methods might be developed and how the underlying philosophy of these methods might be applied to a larger range of problems than hitherto.

REFERENCES

[1] J. M. Abbott, Instability in Quasi-Newton methods for the optimisation of nonlinear functions, M.Sc.Dissertation, University of Essex, 1971.

[2] Y. Bard, On a numerical instability of Davidon-like methods, Math. Comp., 22 (1968) 665-666.

[3] J.G.P. Barnes, An algorithm for solving non-linear equations based on the secant method, Comput. J., 8 (1965) 66-72.

[4] L. Bittner, Eine verallgemeinerung des sekantenverfahrens zur naherungsweisen berechnung der nullstellen eines nichtlinearen gleichungs-systems, Zeit. Tech. Hoch. Dresden, 9 (1959) 325-329

[5] R. P. Brent, IBM Research Report RC3725, (1972).

[6] K. M. Brown and J. E. Dennis, Jr., Derivative free analogues of the Levenberg-Marquardt and Gauss algorithms for nonlinear least squares approximation, Numer. Math., 18 (1972) 289-297.

[7] C. G. Broyden, A class of methods for solving nonlinear simultaneous equations, Math. Comp., 19 (1965) 577-593.

[8] C. G. Broyden, Quasi-Newton methods and their application to function minimisation, Math. Comp., 21 (1967) 368-381.

[9] C. G. Broyden, A new method of solving non-linear simultaneous equations, Comput. J., 12 (1969) 95-100.

[10] C. G. Broyden, The convergence of single-rank quasi-Newton methods, Math. Comp., 24 (1970) 365-382.

[11] C. G. Broyden, The convergence of a class of double-rank minimization algorithms 1, 2, J. Inst. Math. Appl., 6 (1970) 76-90, 222-231.

[12] C. G. Broyden, Recent developments in solving nonlinear simultaneous equations, in Numerical Methods for Nonlinear Algebraic Equations, ed. P. Rabinowitz, Gordon and Breach, London 1970.

[13] C. G. Broyden, The convergence of an algorithm for solving sparse nonlinear systems, Math. Comp., 25 (1971) 285-294.

[14] C. G. Broyden, J. E. Dennis, Jr. and J. J. Moré, On the local and superlinear convergence of quasi-Newton methods, Report TR-72-137, Department of Computer Science, Cornell University 1972.

[15] C. G. Broyden and W. E. Hart, A new algorithm for constrained optimisation, in preparation.

[16] W. C. Davidon, Variable metric method for minimization, A.E.C. Research and Development Report, ANL-5990 (Rev. TID-4500, 14th ed.), 1959.

[17] W. C. Davidon, Variance algorithm for minimization, Comput. J., 10 (1968) 406-410.

[18] J. E. Dennis, On some methods based on Broyden secant approximation to the Hessian, in Numerical Methods for Nonlinear Optimisation, ed. F. A. Lootsma, Academic Press, London, 1972.

[19] L.C.W. Dixon, The choice of step-length, a crucial factor in the performance of variable-metric algorithms, in Numerical Methods for Nonlinear Optimisation, ed. F. A. Lootsma, Academic Press, London, 1972.

[20] R. Fletcher, A new approach to variable metric algorithms, Comput. J., 13 (1970) 317-322.

[21] R. Fletcher and M.J.D. Powell, A rapidly convergent descent method for minimization, Comput. J., 6 (1963) 163-168.

[22] J. L. Greenstadt, Variations of variable-metric methods, Math. Comp., 24 (1970) 1-22.

[23] H. Y. Huang, Unified approach to quadratically convergent algorithms for function minimization, J. Optimization Theory Appl., 5 (1970) 405-423.

[24] H. Y. Huang, J. Levy, Numerical experiments on quadratically convergent algorithms for function minimisation, J. Optimization Theory Appl., 6 (1970) 269-282.

[25] H. Kwakernaak and R.C.W. Strijbos, Extremization of functions with equality constraints, Presented at 7th Mathematical Programming Conference, The Hague.

[26] W. Murray, ed., Numerical Methods for Unconstrained Optimisation, Academic Press, New York, 1972.

[27] B. A. Murtagh and R. W. Sargent, Computational experience with quadratically convergent minimization methods, Computer J., 13 (1970) 185-194.

[28] J. D. Pearson, On variable metric methods for minimization, Comput. J., 12 (1969) 171-178.

[29] M.J.D. Powell, A method for minimizing a sum of squares of non-linear functions without calculating derivatives, Comput. J., 7 (1965) 303-307.

[30] M.J.D. Powell, A hybrid method for nonlinear equations, in Numerical Methods for Nonlinear Algebraic Equations, ed. P. Rabinowitz, Gordon and Breach, London 1970.

[31] M.J.D. Powell, A new algorithm for unconstrained optimisation, in Nonlinear Programming, eds. O. L. Mangasarian, J. B. Rosen and K. Ritter, Academic Press, New York 1970.

[32] M.J.D. Powell, On the convergence of the variable metric algorithms, J. Inst. Math. Appl., 7 (1971) 21-36.

[33] M.J.D. Powell, Some properties of the variable metric algorithm, in Numerical Methods for Nonlinear Optimisation, ed. F. A. Lootsma, Academic Press, London, 1972.

[34] W. C. Rheinboldt, Numerical Solution of Nonlinear Algebraic Systems, CBMS Regional Conference Series in Applied Math., SIAM, Philadelphia, (In preparation).

[35] D. F. Shanno, Conditioning of quasi-Newton methods for function minimization, Math. Comp., 24 (1970) 647-656.

[36] L. K. Schubert, Modification of a quasi-Newton method for nonlinear equations with a sparse Jacobian, Math. Comp., 24 (1970) 27-30.

[37] S.O.W. Soul, The numerical solution of discretized boundary value problems, M.Sc. Dissertation, University of Essex, 1972.

[38] L. Tornheim, Convergence of multi-point iterative methods, J. ACM., 11 (1964) 210-220.

[39] P. Wolfe, The secant method for solving nonlinear equations, Comm. ACM., 2 (1959) 12-13.

COMPUTER ORIENTED ALGORITHMS FOR SOLVING SYSTEMS OF SIMULTANEOUS NONLINEAR ALGEBRAIC EQUATIONS*

Kenneth M. Brown

1. Introduction.

In this paper we consider the problem of solving N real (say transcendental) equations in N real unknowns. Let the system be given as

$$
(1)\qquad
\begin{aligned}
f_1(\underline{x}) &= f_1(x_1,x_2,\ldots,x_N) = 0,\\
f_2(\underline{x}) &= f_2(x_1,x_2,\ldots,x_N) = 0,\\
&\;\;\cdot\quad\cdot\quad\cdot\quad\cdot\quad\cdot\\
f_N(\underline{x}) &= f_N(x_1,x_2,\ldots,x_N) = 0.
\end{aligned}
$$

We write (1) in vector form as $\underline{f}(\underline{x}) = \underline{0}$, and we shall assume throughout that the f_i are continuously

*This work was supported in part by the National Science Foundation under Grant GJ-32552 and in part by the University Computer Center of the University of Minnesota.

differentiable. We shall present what we consider to be the best algorithms for the various facets of the problem of solving (1): getting into a region of local convergence from poor initial estimates; achieving guaranteed convergence to a root from anywhere within a specified region by suitably restricting the functions f_i; using a technique in a vicinity of a root (<u>local technique</u>) which is fast, stable and which does not require the user to furnish derivatives of the functions f_i; and, finally, obtaining additional roots of a nonlinear system without converging again to previously found roots (unless they are in fact multiple roots). We include numerical results and give in the Appendix a FORTRAN IV program for our derivative-free, quadratically convergent local method.

2. Methods for Use with Arbitrary or Poor Initial Guesses.

It sometimes happens that there are no good initial estimates available for the solution of (1), perhaps because the physics or geometry of the problem makes the attainment of good starting guesses untenable. In such cases we want to use methods which bring us systematically, even if slowly, into a vicinity of a root in which one of the faster (local) techniques is guaranteed to work. We mention two such approaches: continuation by differentiation and heuristic search.

Let us define a function $\underline{h}(\underline{x},t)$ such that $\underline{h}(\underline{x},0) \equiv \underline{f}(\underline{x})$ and such that the equation $\underline{h}(\underline{x},1) = \underline{0}$ has a known solution; for example

$$\underline{h}(\underline{x},t) = \underline{f}(\underline{x}) - \underline{f}(\underline{x}^0)\cdot t \qquad \text{(Broyden [9])}.$$

Here of course the "known solution" is $\underline{x} = \underline{x}^0$. Suppose we are trying to solve (1) by an iterative method, but that $\underline{x}^0$ is such a poor starting guess that the application of the method from that starting guess would yield divergence. On the other hand since we have such a good starting guess to the problem $\underline{h}(\underline{x},1-\varepsilon) = \underline{0}$ (where ε is a small positive number) namely $\underline{x}^0$, why not solve a succession of subproblems

$\underline{h}(\underline{x},t_i) = \underline{0}$, with $t_i \to 0$? The ith subproblem should have available to it a good starting guess, namely, the solution to the $i-1^{st}$ subproblem; moreover, the solution, say $\underline{x}^*$, of the final subproblem $\underline{h}(\underline{x},0) = \underline{0}$, is the desired solution of (1) since $\underline{f}(\underline{x}^*) = \underline{0}$ by the definition of $\underline{h}(\underline{x},t)$. This basic idea of systematically reducing the parameter t from 1 to 0 leads naturally to an initial value problem for ordinary differential equations. This approach, continuation by differentiation, is due originally to Davidenko [13]. Further contributions to this idea have been made by Boggs [1], Bosarge [2], Broyden [9], Davis [14] and Meyer [20].

Boggs' approach [1] is especially interesting in that the independent variable of integration is allowed to lie in the half-open interval $[0,\infty)$, rather than being restricted to a finite interval, say [0,1]; such a restriction usually requires that $\underline{x}(1)$ be computed (by numerical integration) to the accuracy desired in $\underline{x}^*$, the solution of $\underline{f}(\underline{x}) = \underline{0}$.

Boggs' technique [1], on the other hand, does not require such accurate numerical solutions to the resulting differential equations; his approach further permits the use of the A-stable integration techniques of Dahlquist [12]. To make these ideas more concrete let us look at Boggs' specific suggestion for the function $\underline{h}(\underline{x},t)$ having the properties that $\underline{h}(\underline{x},0) = \underline{0}$ has a known solution, $\underline{x}^0$, and $\underline{h}(\underline{x},t) \to \underline{f}(\underline{x})$ as $t \to \infty$, namely,

$$\underline{h}(\underline{x},t) = \underline{f}(\underline{x}) - e^{-t}\underline{f}(\underline{x}^0) \qquad \text{(Boggs [1]).} \tag{2}$$

On differentiating $\underline{h}(\underline{x}(t),t) = \underline{0}$ with respect to t, one obtains

$$\underline{x}'(t) = -\underline{h}_{\underline{x}}^{-1}(\underline{x}(t),t)\underline{h}_{t}(\underline{x}(t),t), \tag{3}$$

and as Boggs points out [1, p. 768], for the specific example (2), this last equation, (3), reduces to the initial value problem

$$\underline{x}' = -J^{-1}(\underline{x})\underline{f}(\underline{x}), \qquad \underline{x}(0) = \underline{x}^0. \tag{4}$$

Here $J(\underline{x})$ denotes the Jacobian matrix of the function $\underline{f}$ at the point $\underline{x}$, that is, the matrix whose i^{th}, j^{th} entry is $\partial f_i/\partial x_j$ evaluated at $\underline{x}$. It is assumed of course that $J(\underline{x})$ is nonsingular in a suitable region. The crucial thing to observe is that a solution of (4) also satisfies $\underline{h}(\underline{x},t) = \underline{0}$, so that from (2), $\underline{f}(\underline{x}(t)) = e^{-t}\underline{f}(\underline{x}^0)$ which implies that $\underline{x}(t) \to \underline{x}^*$ as $t \to \infty$. The initial value problem (4) is the basic equation for which Boggs [1] develops a predictor-corrector algorithm. He then proposes modifications of this algorithm which are efficient computationally. Incidentally, as Boggs notes [1, p. 768], applying Euler's method with a step size of one to the integration of (4) is equivalent to doing a single step of Newton's method when solving $\underline{f}(\underline{x}) = \underline{0}$; i.e.,

$$\underline{x}^1 = \underline{x}^0 - J^{-1}(\underline{x}^0)\underline{f}(\underline{x}^0).$$

Heuristic search procedures are used when there is little information available about the derivatives of the functions, f_i, or, for example, when some of these derivatives don't even exist. The basic idea here is to form the nonnegative function $\phi = \Sigma f_i^2$, and look at those subregions of E^N in which $\phi(x_1,\ldots,x_N)$ is unimodal. One then tries a strategy for changing the parameters x_i and observes its effect on the magnitude of ϕ; such a strategy is called a search procedure. An optimal strategy is one which achieves the greatest reduction in ϕ while requiring the minimum number of evaluations of the f_i. There are no foolproof or even "usually works" algorithms in this category; however, an adaptive step size random search technique due to Schumer and Steiglitz [29] appears to be reasonable for multidimensional problems which do not contain narrow valleys or ridges [29, p. 274]. This technique is developed by first looking at a random search procedure (for a hyperspherical surface) which uses optimum step lengths at each point. The optimum step length technique is then modeled by an adaptive method which approximates its behavior.

3. Some Methods Which Are Guaranteed To Converge.

It is logical to ask whether, by demanding considerable structure on our function $\underline{f}$, we might be able to construct algorithms which are not so sensitive to starting guess locations and which will in fact converge for all $\underline{x}^0$ inside some suitable (and not ridiculously small) region. The answer is, happily, yes and we shall look at two such techniques here: Schechter's nonlinear successive overrelaxation process [28] and a combinatorial search algorithm based on a constructive proof of the Brouwer fixed-point theorem due to Scarf [27] and Kuhn [19].

There has been a considerable amount of work done on the nonlinear optimization problem: find the points which minimize (or maximize) the real valued function $G(x_1,x_2,\ldots,x_N)$. See, for example, Chapters 8 and 14 of the Ortega-Rheinboldt book [22]. A special case of this problem is

$$\text{(5)} \qquad \underset{\underline{x}}{\text{minimize}}\ G(\underline{x}) \text{ where } G(\underline{x}) \equiv \sum_{i=1}^{N} f_i^{\,2}(\underline{x}),$$

and where the f_i are the nonlinear functions of (1). In this instance $G(\underline{x})$ takes on the minimum value zero at all solutions of the nonlinear system (1). One approach to solving (5) is to begin with a starting guess $\underline{x}^0$ and select a direction $\underline{d}^0$ such that $G(\underline{x}) < G(\underline{x}^0)$ for vectors $\underline{x}$ of the form $\underline{x}^0 + w\underline{d}^0$ for sufficiently small positive values of w. Let grad $G(\underline{x})$, called the <u>gradient vector</u> of G evaluated at $\underline{x}$, be defined as

$$\text{grad } G(\underline{x}) \equiv (\partial G/\partial x_1, \partial G/\partial x_2, \ldots, \partial G/\partial x_N)^T.$$

It can be shown that the direction - grad $G(\underline{x})$ is the one in which $G(\underline{x})$ decreases most rapidly (locally). It is also easy to show that grad $G(\underline{x}) = 2J^T(\underline{x}) \cdot \underline{f}(\underline{x})$, where, as before, $J(\underline{x})$ denotes the Jacobian matrix of the function vector $\underline{f}$. Thus, the class of <u>gradient methods</u> has the form

$$(6) \qquad \underline{x}^{p+1} = \underline{x}^p - wJ^T(\underline{x}^p) \cdot \underline{f}(x^p), \quad p = 0,1,2,\ldots$$

where $\underline{x}^0$ is given. Here we have absorbed the factor 2 into the parameter w. Gradient techniques have enjoyed some popularity historically, but have not proven as successful in practice on structured problems as some techniques given below.

Another approach, due to Schechter, is essentially a generalization of the technique of successive overrelaxation (SOR) from linear systems to their nonlinear counterparts. Given a sequence of indices $\{i_p\}$ which exhaust the set, Z_N, of the first N natural numbers infinitely often, and a sequence of numbers $\{w_p\}$, $p = 0,1,2,\ldots$, and a starting guess $\underline{x}^0$, Schechter's <u>approximate</u> <u>relaxation</u> process is defined by

$$(7) \quad \begin{cases} x_k^{p+1} = x_k^p + w_p h_p, & k = i_p \\ x_k^{p+1} = x_k^p, & k \neq i_p \end{cases}$$

where $$h_p = -f_k(\underline{x}^p)/f_{kk}(\underline{x}^p)$$

and where, as usual, $f_{kk} \equiv \partial f_k/\partial x_k$. Schechter has proven convergence and uniqueness for the case in which the function $\underline{f}$ from (1) has a positive definite

Jacobian matrix and at the same time represents the gradient of a real-valued, twice-continuously differentiable function G which is uniformly bounded on some convex domain.

THEOREM [28, p. 184]. Let $G: E^N \to E$, $G \in C^2(E^N)$, and for all $x \in E^N$, let $G(\underline{x}) \leq m$ for some constant m. Further, let grad $G(\underline{x}) = \underline{f}(\underline{x})$ with $\underline{f}: E^N \to E^N$ and let the Jacobian matrix of $\underline{f}$, $J(\underline{x})$, be positive definite. Assume that the level set $K = \{\underline{x}: G(\underline{x}) \leq G(\underline{x}^0)\}$ is bounded for some $\underline{x}^0 \in E^N$. Finally let

$$a_i = \min_{\underline{x} \in K} f_{ii}(\underline{x}) \qquad (i \in Z_N)$$

and assume there exist positive constants b_i such that $f_{ii}(\underline{x}) \leq b_i$ for all $\underline{x} \in E^N$. Then for a suitable choice of $\{w_p\}$, the approximate relaxation process (7) converges to a unique solution of (1).

In order to determine an acceptable sequence $\{w_p\}$, Schechter sets $c_0 = \min_i (a_i/b_i)$ and lets c be chosen such that $0 < c < c_0 \leq 1$. He proves that if the w_p satisfy $0 < c \leq w_p \leq 2c_0 - c$, $p = 0,1,2,\ldots$, then the $G(\underline{x}^p)$ are monotone nonincreasing, all the x^p are in K, and $f_{i_p}(\underline{x}^p) \to 0$.

Schechter's technique can be used when certain nonlinear elliptic boundary value problems are solved by finite-difference techniques. Concus [10] has applied Schechter's method to the solving of the Euler equation for the Plateau problem in restricted form.

A truly elegant computational method for finding the fixed points of a mapping has been given by Scarf [27] and refined by Kuhn [19]. Brouwer's fixed-point theorem states that a continuous mapping of a simplex into itself must have at least one fixed point. Scarf [27] has given a constructive proof of this theorem, the construction being remarkably easy to implement on a digital computer. For functions satisfying the hypotheses of Brouwer's theorem, almost arbitrary starting guesses can be used and the method proceeds automatically to the solution. Kuhn [19] has refined Scarf's original work. The procedure has already been used for a number of important applications; e.g., a pure trade model of the economy gives rise to a mapping function, $\underline{f}$, satisfying the Brouwer hypotheses. The strength of this approach is in being able to get into a region of local convergence from poor initial guesses; the weakness is that it may take as much time to obtain a second significant digit of accuracy in the solution as it did to arrive at the first significant digit. What is needed, of course, is a fast local technique to switch over to -- once we are close enough to the root. One such technique is Newton's method; an even better technique is given in the next section.

4. A Rapidly Convergent Local Technique -- Brown's Method.

Let us recall for a moment the derivation of Newton's method. First the functions (1) are expanded *simultaneously* (all at the same time) about a point $\underline{x}^n$ assumed to be close to the solution, $\underline{x}^*$, yielding

$$(8) \qquad \underline{f}(\underline{x}) = \underline{f}(\underline{x}^n) + J(\underline{x}^n)\cdot(\underline{x}-\underline{x}^n) + \text{H. O. T.},$$

where as usual we take points $\underline{x}$ close enough to $\underline{x}^n$ so that the higher order terms (H. O. T.) are negligible when compared to the term involving $(\underline{x}-\underline{x}^n)$. We have also assumed that we are close enough to the solution so that $f(\underline{x}) \sim \underline{0}$, or $\underline{0} \sim \underline{f}(\underline{x}^n) + J(\underline{x}^n)\cdot(\underline{x}-\underline{x}^n)$. Solving for $\underline{x}$ gives us

$$(9) \qquad \underline{x}^{n+1} \equiv \underline{x} = \underline{x}^n - J^{-1}(\underline{x}^n)\cdot\underline{f}(\underline{x}^n).$$

The point $\underline{x}$ thus obtained is renamed $\underline{x}^{n+1}$. Beginning with a starting guess, $\underline{x}^0$, one can solve (9) iteratively for $n = 0,1,2,\ldots$. This is Newton's method. Obviously for (9) to be well-defined $J(\underline{x})$ must be nonsingular in a suitable neighborhood of $\underline{x}^*$. A derivative-free analogue of Newton's method, often called the discrete Newton's method is obtained from (9) by replacing the ith, j^{th} entry of the Jacobian matrix, namely $\partial f_i/\partial x_j$, by the first difference quotient approximation

$$(10) \qquad \frac{f_i(\underline{x}^n + h^n\underline{e}_j) - f_i(\underline{x}^n)}{h^n} .$$

Here $\underline{e}_j$ denotes the jth unit vector and the scalar h^n is normally chosen such that $h^n = 0(||\underline{f}(\underline{x}^n)||)$; with this choice it can be proven (Dennis [15] and Samanskii [26]) that the discrete Newton's method has second order convergence, the same rate of convergence as the ordinary Newton's method. To give some feel for how fast "second order convergence" is, let us observe that for problems which are properly normalized, second order convergence implies roughly a doubling of the number of significant digits in the approximate solution per each iteration: quite fast indeed. Thus we can go from an approximation accurate to just one digit to an approximation accurate to 16 digits in only four applications of (9).

Brown [3,4,5,6] proposed a local method which handles the functions (1) <u>one at a time</u> so that information obtained from working with f_1 can be incorporated when working with f_2, etc. A successive substitution scheme is used rather than the simultaneous treatment of the f_i characteristic of Newton's method. Brown's method is derivative-free; moreover, second order convergence has been proven by Brown and Dennis [7]. The method consists of applying the following steps:

STEP 1. Let $\underline{x}^n$ denote an approximation to the solution $\underline{x}^*$ of (1). Expand the first function f_1 in an approximate Taylor series expansion about the point $\underline{x}^n$. By "approximate" we mean an expansion in which the actual (analytic) partial derivatives are replaced by first difference quotient approximations; i.e., ignoring higher order terms,

$$(11) \qquad f_1(\underline{x}) \sim f_1(\underline{x}^n) + f_{1x_1;h}(\underline{x}^n)\cdot(x_1-x_1{}^n) + f_{1x_2;h}(\underline{x}^n)\cdot(x_2-x_2{}^n)+\ldots+f_{1x_N;h}(\underline{x}^n)\cdot(x_N-x_N{}^n).$$

Here $f_{1x_j;h}(\underline{x}^n)$ is defined to be the first difference quotient approximation (10) with $i = 1$. If $\underline{x}^n$ is close enough to $\underline{x}^*$, $f_1(\underline{x}) \sim 0$, and we can equate (11) to zero and solve for that variable, say x_N, whose corresponding approximate partial derivative, $f_{1x_N;h}(\underline{x}^n)$, is largest in absolute value. This gives

$$(12) \qquad x_N = x_N^n - \sum_{j=1}^{N-1} (f^n_{1x_j;h}/f^n_{1x_N;h}) \cdot (x_j - x_j^n) - f_1^n / f^n_{1x_N;h},$$

where $f^n_{1x_j;h} \equiv f_{1x_j;h}(\underline{x}^n)$, as given in (10), and $f_1^n \equiv f_1(\underline{x}^n)$. The constants $f^n_{1x_j;h}/f^n_{1x_N;h}$ are saved (stored) -- in the computer implementation of the algorithm -- for future use. Brown [5, p. 564] has shown that under the usual hypotheses for Newton's method there will always be at least one non-zero partial derivative, and, of course, the corresponding approximate partial (10) will also be nonvanishing. Thus the solution procedure (12) will be well defined. By choosing the approximate partial derivative of largest absolute value to divide by, a partial pivoting effect is achieved similar to what is often done when using the Gaussian elimination process for solving linear systems. This enhances the numerical stability of the method. We observe from (12) that

x_N is a linear function of the N-1 variables $x_1, x_2, \ldots, x_{N-1}$, and for purposes of clarity later we rename the left-hand side of (12) as $L_N(x_1, x_2, \ldots, x_{N-1})$ and define $L_N^n \equiv L_N(x_1^n, \ldots, x_{N-1}^n)$.

STEP 2. Define a new function g_2 of the N-1 variables $x_1, \ldots, x_{N-1}$ which is related to the second function, f_2, of (1) as follows

$$(13) \qquad g_2(x_1, \ldots, x_{N-1}) \equiv f_2(x_1, \ldots, x_{N-1}, L_N(x_1, \ldots, x_{N-1})).$$

Let g_2^n be defined as $g_2^n \equiv f_2(x_1^n, \ldots, x_{N-1}^n, L_N^n)$. Now expand g_2 in an approximate Taylor series expansion about the point $(x_1^n, \ldots, x_{N-1}^n)$, linearize (ignore higher order terms) and solve for that variable, say x_{N-1}, whose corresponding approximate partial derivative, $g_{2x_{N-1};h}$, is largest in magnitude:

$$(14) \qquad x_{N-1} = \left\{ x_{N-1}^n - \sum_{j=1}^{N-2} (g_{2x_j;h}^n / g_{2x_{N-1};h}^n) \cdot (x_j - x_j^n) - g_2^n / g_{2x_{N-1};h}^n \right\}$$

Here the approximate partial derivative $g^n_{2x_j;h}$ is given by

$$(15) \qquad g^n_{2x_j;h} \equiv \frac{g_2(x^n_1,\ldots,x^n_{j-1},x^n_j+h^n,x^n_{j+1},\ldots,x^n_{N-1}) - g^n_2}{h^n}$$

where h^n is a small positive number. In the convergence theorem given below we shall show how to choose h^n so as to guarantee second order convergence. Again let us note from (14) that x_{N-1} is a linear function of the remaining N-2 variables and let us denote that linear function (i.e., the right hand side of (14)) by $L_{N-1}(x_1,\ldots,x_{N-2})$. Again the ratios formed, $g^n_{2x_j;h}/g^n_{2x_{N-1};h}$, $j = 1,\ldots,N-2$, and $g^n_2/g^n_{2x_{N-1};h}$, should be stored for future use in any computer implementation of this algorithm.

STEP 3. Define

$$(16) \qquad g_3(x_1,\ldots,x_{N-2}) \equiv f_3(x_1,\ldots,x_{N-2},L_{N-1},L_N)$$

with the argument of L_{N-1} being $(x_1,\ldots,x_{N-2})$ and (please note) the argument of L_N being

(17) $$(x_1, x_2, \ldots, x_{N-2}, L_{N-1}(x_1, x_2, \ldots, x_{N-2})).$$

We repeat the basic process of 1) approximate Taylor expansion of the function g_3 about $(x_1^n, \ldots, x_{N-2}^n)$, followed by 2) linearization of the resulting expansion, followed by 3) equating to zero and solving for one variable, say x_{N-2} (whose corresponding approximate partial derivative $g_{3x_{N-2};h}$ is largest in magnitude) as a linear combination, L_{N-2}, of the now remaining N-3 variables.

We continue in this fashion, eliminating one variable for each equation treated. Every time we obtain a new linear expression, L_{N-k}, for one of the variables, say x_{N-k}, in terms of the remaining N-k-1 variables, $x_1, x_2, \ldots, x_{N-k-2}, x_{N-k-1}$, we use this linear expression wherever x_{N-k} had appeared in the previously defined linear expressions $L_{N-k+1}, L_{N-k+2}, \ldots, L_{N-1}, L_N$. Looked at another way, each step in the algorithm adds one more linear expression to a linear system. During the k + 1st step of the algorithm, it is necessary to evaluate g_{k+1}, i.e., f_{k+1}, for various arguments. The values of the last k components of the argument of f_{k+1} are obtained by back-substitution in the linear system $L_N, L_{N-1}, \ldots, L_{N-k+1}$ which has been built up. The points which are back-substituted consist of the point $(x_1^n, \ldots, x_{N-k}^n) \equiv X_{N-k}^n$ together with the points $X_{N-k}^n + h^n \underline{e}_j$, $j = 1, \ldots, N-k$, where $\underline{e}_j$ denotes the jth unit vector. These arguments are required to determine the quantities g_{k+1}^n and $g_{k+1,j;h}^n$, $j = 1, \ldots, N-k$, needed for the elimination of the k+1st variable, say x_{N-k}, by the basic processes of expansion, linearization and solution of the resulting expression. The process results, for each k, in the k+1st variable, say x_{N-k}, being expressed as a linear combination, L_{N-k}, of the remaining N-k-1 variables.

STEP N. At this stage we have $g_N \equiv f_N(x_1, L_2, L_3, \ldots, L_N)$ where the L_j's are obtained by back-substitution in the N-1 rowed triangular linear system which now has the form

$$L_i = x_i^n - \sum_{j=1}^{i-1} (g^n_{N-i+1,x_j;h} / g^n_{N-i+1,x_i;h}) \cdot (L_j - x_j^n) - g^n_{N-i+1} / g^n_{N-i+1,x_i;h}, \quad i = N, N-1, \ldots, 2, \tag{18}$$

(with $g_1 \equiv f_1$ and $L_1 \equiv x_1$) so that g_N is just a function of the single variable x_1. Now expanding, linearizing and solving for x_1, we obtain

$$x_1 = x_1^n - g_N^n / g^n_{N,x_1;h}. \tag{19}$$

We use the point x_1 thus obtained as the next approximation, x_1^{n+1}, to the first component, x_1^*, of the solution vector $\underline{x}^*$. We rename x_1 as L_1 and back-solve the L_j system (18) to get improved approximations to the other components of $\underline{x}^*$. Here we take as x_j^{n+1} the value obtained for L_j when back-solving (18).

The "successive substitution" nature of the algorithm allows the most recent information available to be used in the construction of the next function argument, similar to what is done in the Gauss-Seidel process for linear [11, pp. 166-169] and nonlinear [23] systems of equations.

REMARK. As is easy to show by example [5, p. 563], Brown's method is not mathematically equivalent to Newton's method.

5. A Local, Second Order Convergence Theorem for Brown's Method [7, pp. 10-11].

The weak hypothesis.

Let $\underline{x}^*$ be a zero of the function vector $\underline{f}$ of (1) and let the Jacobian of $\underline{f}$ be continuous in $\bar{S}(\underline{x}^*;R) \equiv \{\underline{x} \in E^N: ||\underline{x}-\underline{x}^*||_\infty \leq R\}$ and be nonsingular at $\underline{x}^*$, where $R > 0$.

The strong hypothesis.

Let $K \geq 0$ and assume that, in addition to the weak hypothesis, $\underline{f}$ satisfies the property that

$$||J(\underline{x})-J(\underline{x}^*)||_\infty \leq K||\underline{x}-\underline{x}^*||_\infty, \text{ for } ||\underline{x}-\underline{x}^*||_\infty \leq R.$$

THEOREM. [7, p. 10]. If $\underline{f}$ satisfies the weak hypothesis then there exist positive numbers, r,

ε such that if $\underline{x}^0 \varepsilon S(\underline{x}^*;r)$ and $\{h^n\}$ is bounded in modulus by ε, Brown's method for nonlinear systems applied to $\underline{f}$ generates a sequence $\{\underline{x}^n\}$ which converges to $\underline{x}^*$. Moreover, if $\underline{f}$ satisfies the strong hypothesis and $\{h^n\}$ is $O(\{|f_1(\underline{x}^n)|\})$, then the convergence is at least second order.

The (rather lengthy) proof is given in [7, pp. 11-27].

6. The Choice of h^n in the Computer Implementation of Brown's Method.

Recall that the condition for second order convergence for Newton's method was $\{h^n\} = O(\{||\underline{f}(\underline{x}^n)||\})$ [15], [26]. The requirement in the above theorem is more stringent than this and the $||\underline{f}(\underline{x}^n)||$ requirement would certainly suffice in the convergence analysis; however, we use the $|f_1(\underline{x}^n)|$ requirement because it is easy to implement computationally. In the actual computer program (the FORTRAN IV subroutine NONLIN given in the appendix of this paper) we must temper our choice of h^n even further. Suppose, for example, that $||\underline{x}^n||_\infty = 0.001$ but $|f_1(\underline{x}^n)| = 1000$; this would provide poor approximations to the partial derivatives. We wish to prevent the size of h^n from being absurd when compared to the magnitude of $||\underline{x}^n||_\infty$, while at the same time satisfying the conditions of the theorem asymptotically (i.e. as the solution, $\underline{x}^*$, is approached). The following strategy is used for choosing the amount, h_j^n, by which to increment x_j^n when working with the ith function, f_i, (see (10)):

$$h_j^n = \max\{\alpha_{ij}^n; \; 5 \times 10^{-\beta+2}\}$$

where

$$\alpha_{ij}^n = \min\{\max(|f_1^n|, |g_2^n|, \ldots, |g_i^n|); \; .001 \times |x_j^n|\}$$

where β is the machine tolerance, i.e., the number of significant digits carried by the machine.

7. Computational Efficiency and Numerical Results for Brown's Method.

If we count the number of function values of the f_i needed per iteration of Brown's method, we see from §4 that the first step requires N+1 evaluations of f_1, the second step N evaluations of f_2, the third N-1 evaluations of f_3, etc., so that the total is

$$\sum_{i=2}^{N+1} i = \frac{N^2 + 3N}{2}.$$

This compares quite favorably with the discrete Newton's method which requires N^2+N individual function component evaluations per iterative step; moreover, Newton's method requires N^2 partial derivative evaluations and N function component evaluations per iterative step. There is a corresponding savings in storage locations required -- from N^2+N for Newton's method or the discrete Newton's method to $N^2/2 + 3N/2$ locations for Brown's method. It must be stressed here that the savings in function values applies only to saving function values of the f_i of the original system (1), for Brown's method adds a number of other functions to be evaluated, namely the linear functions, L_k. The evaluations of the L_k are not included in the count above, "$N^2/2 + 3N/2$". Thus Brown's method will only effect a real savings of total computational effort vs. Newton's methods (usual and discrete) when the functions f_i are expensive to evaluate in terms of amount of computation. On the other hand, the method seems to be extremely stable locally in practice and has been used very successfully over the last six years on more than 100 different problems, even in cases where the f_i are "cheap" to evaluate. A somewhat lengthy and tedious computation reveals that the break-even point as regards computational efficiency is for systems (1) whose average equation entails more than N^2 multiplications. Many systems arising from physical problems are at least this complicated; for some systems each function value f_i may require the solving of a differential equation -- making such systems ideal candidates for Brown's method.

Because Brown's method works with one equation, $f_i = 0$, at a time and uses information thus obtained immediately when dealing with the next equation $f_{i+1} = 0$, there is an optimal ordering strategy for the system of equations (1). The equations should be preordered so that the linear ones, or most nearly linear ones, come first and then the equations become progressively more nonlinear (as say measured by their degree). Such a strategy does no good for Newton's method or for the discrete Newton's method because all the equations, $f_i = 0$, $i = 1,\ldots,N$, are treated simultaneously by those methods. The first numerical example illustrates this contrast quite dramatically.

EXAMPLE 1. [5, p. 567]. Consider the extreme case of the situation alluded to in the previous paragraph, namely a system in which every equation is linear except for the very last one, with that final equation being highly nonlinear:

$$\begin{cases} f_i(\underline{x}) \equiv -(N+1) + 2x_i + \sum_{\substack{j=1 \\ j\neq i}}^{N} x_j, \quad i = 1,\ldots,N-1, \\ f_N(\underline{x}) \equiv -1 + \prod_{j=1}^{N} x_j. \end{cases}$$

The problem was run for N = 5, 10 and 30. The starting guess for all cases was a vector having 0.5 in each component. The results are given in Table 1 below. We note that Brown's method converged in each case to the root, a vector all of whose components are 1.0. For N = 5 Newton's method converged to the root given approximately by (-.579, -.579, -.579, -.579, 8.90); however, for N = 10 and 30 Newton's method diverged quite rapidly. As Table 1 shows, even after just one iteration the components of $\underline{x}^1$ are quite large. The failure of Newton's method on this problem is not attributable to singularities of the Jacobian matrix, since the Jacobian matrix is nonsingular at the starting guess and at the two roots. In the table "diverged" means that $||\underline{x}^n||_\infty \to \infty$ (the program was terminated whenever $||\underline{x}^n||_\infty > 10^{35}$ whereas "converged" means that each component of $\underline{x}^{n+1}$

agreed with the corresponding component of $\underline{x}^n$ to 15 significant digits and $||\underline{f}(\underline{x}^{n+1})||_{\ell_2} < 10^{-15}$.

Table 1

Computer results for Example 1

N	Newton's Method	Brown's Method
5	converged in 18 its.	converged in 6 its.
10	diverged, $\|\|\underline{x}^1\|\|_{\ell_2} \sim 10^3$	converged in 7 its.
30	diverged, $\|\|\underline{x}^1\|\|_{\ell_2} > 10^6$	converged in 9 its.

Even for problems in which one or more of the functions is only approximately linear at the root, Brown's method capitalizes on the preordering strategy above, whereas Newton's method does not -- as the next example shows.

EXAMPLE 2. [5, p. 567]. In the following system, $f_1(x_1,x_2)$ is approximately linear near the roots:

$$\begin{cases} f_1(x_1,x_2) = x_1^2 - x_2 - 1 \\ f_2(x_1,x_2) = (x_1-2)^2 + (x_2-0.5)^2 - 1. \end{cases}$$

The system has roots at

$$\underline{r} \sim (1.54634288,\ 1.39117631) \quad \text{and}$$

$$\underline{s} \sim (1.06734609,\ 0.139227667).$$

The starting guess used was (0.1,2.0). The computer results are given in Table 2. The same convergence test was used as in Example 1.

Table 2

Computer results for Example 2

Method	Result
Newton	converged to $\underline{s}$ in 24 iterations
Brown	converged to $\underline{s}$ in 10 iterations

REMARK. In order to compute the total number of evaluations of the individual function components

f_i necessary to achieve convergence in Examples 1 and 2, simply multiply the number of iterations given in the tables by $N^2 + N$ for Newton's method and by $(N^2 + 3N)/2$ for Brown's method. As before, N denotes the order of the system.

In some quite recent (Spring, 1972) work Möhlen [21] has compared Powell's new (1968-69) method [24], [25] with Brown's method for solving systems of simultaneous nonlinear equations. Möhlen has gone on to propose modifications of both of these algorithms which attempt to make them more robust (capable of converging in a larger region); this is achieved in some cases, but at the expense of increased computational effort, a greater number of function component evaluations. I feel that Möhlen's work is in the right direction: modifications which make local algorithms more global are quite desirable; however, I also feel that there is room for considerable research on the precise nature of such modifications and that we are still a long way from an optimally modified local method.

The program which Möhlen used to implement Brown's method was based on an older form of the method [4] which was only roughly second order and did not include the h^n selection strategy given in §6 and used in the current program (NONLIN, the FORTRAN IV subroutine given in the appendix). Recall that the current h^n selection procedure guarantees at least second order convergence. Using the current program results in a savings of one or sometimes two iterations, i.e., of $(N^2 + 3N)/2$ or $N^2 + 3N$ individual f_i evaluations in the figures quoted below from Möhlen's paper [21]; however, we quote Möhlen's results rather than our own for these examples, as it is always better to have an outside, independent verification of results claimed for one's own method (this is similar to having an outside accounting firm audit the books of a large company as opposed to letting the company prepare its own annual financial report!).

EXAMPLE 3. This example was taken from

Powell's own paper [24, p. 36] and is reported by Möhlen [21, p. 51]:

$$\begin{cases} f_1(x_1,x_2) = 10000\ x_1x_2 - 1 \\ f_2(x_1,x_2) = \exp(-x_1) + \exp(-x_2) - 1.0001. \end{cases}$$

The solution is given approximately by $\underline{x}^* \sim (.1099\times10^{-4}, 9.096)$. The starting guess used was $\underline{x}^0 = (0,1)$. The results are given in Table 3 below. The convergence (stopping) criterion used was $\Sigma f_i^2 < 10^{-10}$.

Table 3

Computer results for Example 3

Method	Number of Individual Function Component Evaluations Necessary for Convergence
Powell	414
Brown	55

EXAMPLE 4. A problem called "Chebyquad" was

investigated by Fletcher [16, p. 36]. The problem consists of calculating the abscissas for the Chebyshev quadrature rules for N = 2,3,...,7,9. The case N = 8 has no solution. The functions f_i are defined by [18, pp. 345-351]

$$f_i(x_1,\dots,x_N) = \int_0^1 T_i(z)dz - \frac{1}{N}\sum_{j=1}^{N} T_i(x_j),$$

$$i = 1,\dots,N$$

where

$$T_0(z) \equiv 1$$

$$T_1(z) \equiv 2z - 1$$

$$T_{i+1}(z) = 2\cdot(2z-1)\cdot T_i(z) - T_{i-1}(z),$$

$$i = 1,2,\dots\ .$$

It can be shown that

$$\int_0^1 T_i(z)dz = \begin{cases} 0 & , \text{ if } i \text{ is odd} \\ \dfrac{-1}{i^2-1}, & \text{if } i \text{ is even.} \end{cases}$$

In Table 4 below, we quote the results obtained by Möhlen for this problem [21, p. 53]. The starting guess used was $x_i^0 = i/(N+1)$, $i = 1,\ldots,N$ and the convergence criterion used was $\Sigma f_i^2 < 10^{-8}$.

Table 4

Computer results for Example 4

Dimension N	Number of Individual Function Evaluations Required for the Method Indicated	
	Powell	Brown
2	14	15
3	27	27
4	56	42
5	65	60
6	204	108
7	140	105
9	360	270

It is most interesting to note that when Möhlen tried running both methods for the "no solution" case N = 8, Powell's method converged to a stationary point, whereas Brown's method refused to converge to anything.

Of all the test cases studied by Möhlen there was only one system for which Powell's method was superior to Brown's. This system is given in Example 5 below.

EXAMPLE 5 [21, pp. 43, 54-55] and [24, p. 33].

$$f_i(\underline{x}) = \sum_{j=1}^{N} [A_{ij} \cdot \sin(x_j) + B_{ij} \cdot \cos(x_j)] - E_i,$$

$$i = 1, \ldots, N,$$

where the A_{ij} and B_{ij} are random numbers between -100 and +100. The E_i are determined as follows: a point $\underline{x}^*$ is chosen such that each component, x_j^*, of $\underline{x}^*$ is a random number between $-\pi$ and $+\pi$; then $f_i(\underline{x}^*) = 0$ is solved for E_i. The starting guess used, $\underline{x}^0$, is then formed as $x_i^0 = x_i^* + .1\eta_i$, $i = 1, \ldots, N$, where each η_i is itself a random number in the interval $[-\pi, +\pi]$. The convergence test used was $\Sigma f_i^2 < 10^{-3}$. The results obtained by Möhlen [21, p. 55] are given in Table 5 below. In that table the numbers again denote the numbers of individual function component evaluations necessary to achieve convergence. The symbol "-" in the table means that convergence was not obtained within 30 iterations.

REMARK. It may very well turn out that Powell's method is superior to Brown's method for the problem given in Example 5, but several things disturb me about that example:

1) The "randomness" of selecting A_{ij}, B_{ij}, E_i and especially the starting guesses x_i^0. It is fine to choose parameters randomly, as is done in many kinds of simulations, but the results claimed should be averaged over a reasonable number of samples.

2. The fact that only 30 iterations were allowed. I have run a number of examples where the starting guesses were (obviously) not within the region of local convergence for the method, yet the method was stable enough to keep the iterates from "blowing-up"

and eventually (sometimes after as many as 110 iterates) arrive in the region of local convergence. Once there, of course, convergence took place within the next three or four iterations.

3) No report is given about the magnitudes of $\underline{x}^{30}$. It would have been interesting to know whether the approximation $\underline{x}^{30}$ produced by Brown's method was close to the solution or far afield from it.

I am in the process of running NONLIN (the program given in the appendix) on 25 different trials of Example 5 and shall report the results of these experiments in a future paper.

Table 5

Computer results for Example 5

Dimension N	Powell	Brown
5	55	60
	70	60
	55	60
	60	120
	75	80
10	180	260
	180	260
	150	130
	130	130
	210	455
15	405	540
	360	405
	465	945
	405	675
	420	540
20	640	-
	700	-
	680	-
	560	920
	540	690
25	1850	-
	957	-
	1100	-
	1200	-
	2500	-
30	1230	-
	1200	-
	1170	-
	1410	-
	1470	-

8. On Minimizing Computational Effort: The Samanskii Idea.

Consider the following iterative scheme studied by Samanskii [26]:

(20)

$$J(\underline{x}^n;h^n)\cdot\Delta\underline{x}^{n;1} = -\underline{f}(\underline{x}^n), \qquad \underline{x}^{n;1} = \underline{x}^n + \Delta\underline{x}^{n;1}$$

$$J(\underline{x}^n;h^n)\cdot\Delta\underline{x}^{n;2} = -\underline{f}(\underline{x}^{n;1}), \qquad \underline{x}^{n;2} = \underline{x}^{n;1} + \Delta\underline{x}^{n;2},$$

$$J(\underline{x}^n;h^n)\cdot\Delta\underline{x}^{n;3} = -\underline{f}(\underline{x}^{n;2}), \qquad \underline{x}^{n;3} = \underline{x}^{n;2} + \Delta\underline{x}^{n;3},$$

$$\cdot \quad \cdot \quad \cdot \qquad\qquad \cdot \quad \cdot \quad \cdot$$

$$J(\underline{x}^n;h^n)\cdot\Delta\underline{x}^{n;m_n} = -\underline{f}(\underline{x}^{n;m_n-1}), \quad \underline{x}^{n+1} = \underline{x}^{n;m_n-1} + \Delta\underline{x}^{n;m_n}.$$

Here $n = 0,1,2,\ldots,\underline{x}^0$ is given and $J(\underline{x}^n;h^n)$ denotes the discretized Jacobian matrix whose i^{th}, j^{th} entry is given by (10). We note that if $m_0 = m_1 = \ldots = m_n = \ldots = 1$, then (20) reduces to the discrete Newton's method described at the beginning of Section 4. Samanskii has given a convergence theorem [26, p. 134] for the method (20) and has shown that the error at the nth step is given by

$$(21) \qquad ||\underline{x}^n - \underline{x}^*||_\infty \leq C\cdot\beta^p,$$

where C and β are estimated in the hypotheses of Samanskii's theorem and involve bounds on the norms of $\underline{f}$, $J(\underline{x};h)^{-1}$, etc., and where

$$(22) \qquad p = \prod_{i=0}^{n-1} (m_i + 1).$$

From (21) - (22) it is easy to show that the discrete Newton's method has second order convergence, since in that case $m_j = 1$, $j = 0,1,2,\ldots$.

Samanskii poses and solves a most interesting computational question [26, pp. 137-8]. <u>How can the m_i be chosen so as to minimize computational effort?</u> To answer this question, suppose, for example, that $m_i = m$, a constant, for all i. Then the error estimate becomes

$$||\underline{x}^n - \underline{x}^*||_\infty \leq C\cdot\beta^{(m+1)^n}.$$

Now let N_D computations be required in order to construct the discrete Jacobian matrix $J(\underline{x}^n;h^n)$ at each step. Most of these N_D computations will be due to function evaluations of the f_i. Let, moreover, N_S

computations be required to evaluate each $\underline{f}(\underline{x}^{n;j})$ and to solve each linear system produced. Then we require a total of roughly

$$N_T(m) = n(N_D + mN_S) \tag{23}$$

computations in order to produce $\underline{x}^n$. Using $C \cdot \beta^{(m+1)^n}$ as a measure of the accuracy of the approximate solution, $\underline{x}^n$, and letting ε denote the required accuracy, we wish to minimize $N_T(m)$ subject to the constraint that

$$C \cdot \beta^{(m+1)^n} = \varepsilon. \tag{24}$$

On solving (24) for n in terms of m, substituting that expression for n into (23), differentiating and equating to zero, we get the following equation for determining the minimum:

$$\ln(1+m) = 1 + \left(\frac{N_D}{N_S} - 1\right) \cdot \frac{1}{1+m}, \quad m \geq 1.$$

Samanskii has shown that this last equation has a single root. Call it m*. Now take the integer nearest m* as the value of m which will minimize computational effort.

The Samanskii idea for minimizing computational effort can be used for other iterative schemes than (20). To apply this idea to Brown's method, one simply holds the ratios $g^n_{i,x_j;h}/g^n_{i,x_k;h}$ constant for m steps, while recomputing only the function values g^n_i during those steps. Recall that the ratios are the coefficients of the linear system $L_N, L_{N-1}, \ldots, L_1$ built up during each step of the iteration. (See §4.)

9. Deflation Techniques for the Calculation of Further Solutions of a Nonlinear System.

We now consider a class of methods which can be used to find further solutions of (1) -- in addition to those solutions known a priori or found during earlier calculations. The basis for these methods is to define a new system of equations which still has all the remaining solutions of (1), but which does not tend to zero as an old root is approached -- unless that root happens to be a multiple root. This technique is called deflation. The material presented here is based upon the Brown and Gearhart paper [8].

To get a feel for the ideas involved, let us look at the one dimensional situation. Let f be a real valued function of the real variable x. (Most of the results given here also hold for analytic functions of a complex variable.) Let $f(x) = 0$ and suppose that r is a simple solution of that equation, i.e., $f(r) = 0$, but $f'(r) \neq 0$. Now let $s \neq r$ be any additional solution of $f(x) = 0$. We wish to define a function $g(x)$ which has the properties that 1) s will be a solution of $g(x) = 0$, but 2) as $x \to r$, $g(x)$ does not approach zero -- i.e., as the old root is approached, the newly defined function does not tend to zero. A natural choice for $g(x)$ seems to be

$$(25) \quad \begin{cases} g(x) \equiv \dfrac{f(x)}{x-r}, & x \neq r \\[2ex] g(r) \equiv f'(r). & \end{cases}$$

Obviously, $g(s) = 0$; the remaining question is what happens to the values of $g(x)$ as $x \to r$. Now

$$\lim_{x \to r} g(x) = \lim_{x \to r} \frac{f(x) - 0}{x-r}$$

$$= \lim_{x \to r} \frac{f(x) - f(r)}{x-r}$$

$$\equiv f'(r);$$

but by hypothesis r is simple and hence $f'(r) \neq 0$; i.e., $g(x)$ does not approach zero. If we know k simple roots, $r_1, r_2, \ldots, r_k$, and we wish to find r_{k+1} we can extend the definition of g given in (25) as follows:

$$(26) \quad \begin{cases} g(x) \equiv \dfrac{f(x)}{\prod_{i=1}^{k} (x-r_i)}, & x \neq r_i \\[2ex] g(r_i) \equiv \dfrac{f'(r_i)}{(r_i-r_1)\ldots(r_i-r_{i-1})(r_i-r_{i+1})\ldots(r_i-r_k)}, & \\[2ex] \text{for } i = 1,\ldots,k. & \end{cases}$$

Again, the function g defined by (26) has the desired properties that 1) $g(r_{k+1}) = 0$ if r_{k+1} is a zero of f and 2) g(x) does not approach zero as $x \to r_i$, $i = 1,\ldots,k$.

REMARK. Suppose we wish to use Newton's method (as it is given in the one dimensional case):

$$x^{n+1} = x^n - \frac{g(x^n)}{g'(x^n)}, \qquad n = 0,1,2,\ldots,$$

to find a zero of the function g given in (26). At first it seems necessary to compute $g(x^n)$ and $g'(x^n)$ explicitly; however, as has been pointed out in the literature (see, for example, Wilkinson [30, p. 78]),

$$\frac{g'(x)}{g(x)} = \frac{f'(x)}{f(x)} - \sum_{i=1}^{k} \frac{1}{(x-r_i)} . \tag{27}$$

Now the reciprocal of the ratio given in (27) is precisely the value needed when applying Newton's method to the function g; thus the need for computing $g(x^n)$ and $g'(x^n)$ is circumvented.

We now look at generalizations of these ideas (25-26) to higher dimensional problems. As before, let $\underline{f}(\underline{x}) = \underline{0}$ denote the vector form of a system of N real nonlinear equations in N real unknowns. Let E^N denote N-dimensional Euclidean space, with the inner product of $\underline{u},\underline{v} \in E^N$ denoted by $\langle\underline{u},\underline{v}\rangle$, and the associated norm by $||\underline{u}||$. For each $\underline{r} \in E^N$, let $M(\underline{x};\underline{r})$ be a matrix on E^N which is defined for all $\underline{x} \in U_{\underline{r}}$, where $U_{\underline{r}}$ is some open set in E^N such that $\underline{r}$ belongs to the closure of $U_{\underline{r}}$.

DEFINITION. We will say that M is a <u>deflation matrix</u> if for any differentiable function $\underline{f}\colon E^N \to E^N$ such that $\underline{f}(\underline{r}) = \underline{0}$ and $J(\underline{r})$, the Jacobian matrix, is nonsingular, we have

$$\liminf_{i \to \infty} ||M(\underline{x}^i;\underline{r})\cdot\underline{f}(\underline{x}^i)|| > 0 \tag{28}$$

for any sequence $\underline{x}^i \to \underline{r}$, $\underline{x}^i \in U_{\underline{r}}$.

REMARKS. Instead of the Jacobian matrix being nonsingular, we may require, more generally, that the first Frechet derivative, $\underline{f}'(\underline{r})$, be nonsingular. It was also not necessary to require that $\underline{f}$ be defined on all of E^N; however, this was done for convenience.

DEFINITION. The function defined by

$$(29) \qquad \underline{g}(\underline{x}) \equiv M(\underline{x};r)\cdot\underline{f}(\underline{x})$$

is called the deflated function, or the function obtained from $\underline{f}$ by deflating out the simple zero $\underline{x} = \underline{r}$.

We note that if we use an iterative method to find zeros of $\underline{g}$ (i.e., further zeros of $\underline{f}$), we are assured in the sense described by (28) that any sequence of points converging to the simple zero $\underline{r}$ of $\underline{f}$ will not produce a zero of $\underline{g}$.

To deflate out k simple zeros, $\underline{r}_1, \underline{r}_2, \ldots, \underline{r}_k$, we form the deflated function

$$(30) \qquad \underline{g}(\underline{x}) \equiv M(\underline{x};\underline{r}_1)\cdot M(\underline{x};\underline{r}_2)\ldots M(\underline{x};\underline{r}_k)\cdot\underline{f}(\underline{x}).$$

DEFINITION. A matrix $M(\underline{x};\underline{r})$ is said to have Property P if and only if for each $\underline{r} \in E^N$ and any sequence, $\underline{x}^i \to \underline{r}$ with $\underline{x}^i \in U_{\underline{r}}$,

$$\text{whenever } \|\underline{x}^i - \underline{r}\|\cdot M(\underline{x}^i;\underline{r})\ \underline{u}^i \to 0$$

$$\text{for some sequence } \{\underline{u}^i\} \subset E^N,$$

then $\underline{u}^i \to \underline{0}$.

The following lemma is helpful in constructing deflation matrices.

LEMMA. If the matrix $M(\underline{x};\underline{r})$ has Property P, then $M(\underline{x};\underline{r})$ is a deflation matrix. (The proof is given in [8, p. 335].)

Let us now consider two classes of deflations for nonlinear systems. Using the above Lemma it is easy to show that the matrices associated with these classes are deflation matrices.

We call the first class of methods <u>norm deflation</u> and define this class by taking

$$M(\underline{x};\underline{r}_i) = \frac{1}{\|\underline{x}-\underline{r}_i\|} \cdot A,$$

in (30) where A is a nonsingular matrix on E^N. Here any norm can be used. The domain of definition for (30) in this case is given by $E^N - \bigcup_{i=1}^{k} \{\underline{r}_i\}$. In the numerical results given later, we have used $A = I$, the identity matrix. Forsythe and Moler [17, p. 136] have also proposed this special form of norm deflation.

We name the second class of deflation methods as <u>inner product deflation</u>. Here we take $M(\underline{x};\underline{r}_i)$ in (30) to be a diagonal matrix whose j^{th} diagonal element is given by

$$m_{jj} = (\langle \underline{a}_j^{(i)}, \underline{x}-\underline{r}_i \rangle)^{-1}$$

where $\underline{a}_1^{(i)},\ldots,\underline{a}_N^{(i)}$ are nonzero vectors in E^N, $i = 1,\ldots,k$. If we set

$$C(\underline{a}_j) = \{\underline{u} \in E^N\colon \langle \underline{a}_j^{(i)}, \underline{u} \rangle = 0,\ i = 1,\ldots,k\}$$

then the domain of definition for (30) is given by

$$E^N - \bigcup_{i=1}^{k} \bigcup_{j=1}^{N} (C(\underline{a}_j^{(i)}) + \{\underline{r}_i\}).$$

If the $\underline{a}_j^{(i)}$ are chosen as

$$\underline{a}_j^{(i)} = \text{grad } f_j \big|_{x=r_i}$$

we shall refer to the method as <u>gradient deflation</u>. (This choice has proven useful in practice when using Newton's Method.) The jth component of $\underline{g}$ in (30) for gradient deflation is given by

$$(31)\qquad g_j(\underline{x}) = \frac{f_j(\underline{x})}{\prod_{i=1}^{k} \langle \text{grad } f_j(\underline{r}_i), \underline{x}-\underline{r}_i \rangle},$$

$$j = 1,\ldots,N.$$

REMARKS. 1.) The deflation techniques presented here for E^N have the advantage of keeping the iterates away from previously determined roots; however, unlike the one dimensional case, there is no way to define the deflated function $\underline{g}$ for all points at which $\underline{f}$ is defined. 2.) Should (31) be too hard to evaluate for a particular problem, it may be more convenient to use a finite difference approximation to any partial derivative of $\underline{g}$ required by the iterative technique being used; e.g.,

$$\frac{\partial g_i}{\partial x_j} \sim \frac{g_i(\underline{x} + h\underline{e}_j) - g_i(\underline{x})}{h}$$

where g_i denotes the ith component function of the vector function $\underline{g}$, where $\underline{e}_j$ is the jth unit vector and h is a small positive increment which may depend on $\underline{x}$. In the numerical results reported in [8, pp. 338-342], we have used these first differences to approximate the partial derivatives needed.

The definition of the deflation matrix indicates the behavior of the matrix with regard to only simple zeros. We will consider now the effect of the deflation matrix on multiple zeros of $\underline{f}: E^N \to E^N$. For convenience, we will assume that $\underline{f}$ is infinitely differentiable and let $\underline{f}^{(k)}$ denote the kth Frechet derivative of $\underline{f}$ (thus, $\underline{f}^{(k)}(\underline{x}) \cdot v^{(k)}$ denotes the kth derivative of $\underline{f}$ at $\underline{x} \in E^N$ applied k times to the vector $\underline{v} \in E^N$).

DEFINITION. If there is a unit vector $\underline{v} \in E^N$ such that $\underline{f}^{(h)}(\underline{r}) \cdot \underline{v}^{(h)} = \underline{0}$ for all $h = 0,1,2,\ldots,$ then we will say that $\underline{f}$ is flat at $\underline{x} = \underline{r}$.

In E^1 it is easy to show that if we apply the deflation $1/(x-r)$ to $f: E^1 \to E^1$ then provided f is not flat at $x = r$, there is an integer k such that $f(x)/(x-r)^k$ does not have a zero at $x = r$. In performing the numerical calculations, it is not necessary to know the multiplicity of a given zero in order to construct the deflated function. Indeed, when a zero, say $x = r$, is first computed, it suffices to merely divide by the factor $(x-r)$. Each time this zero is computed again (if at all), we simply divide again by the factor $(x-r)$. Thus, unless f is flat at $x = r$, we will ultimately arrive at a deflated function which does not have a zero at r. In higher dimensions, we have

THEOREM. Suppose $\underline{f}: E^N \to E^N$ is infinitely differentiable and $\underline{f}(\underline{r}) = \underline{0}$. Let $M(\underline{x};\underline{r})$ have the property (P) (in particular, M is a deflation matrix). Then, provided $\underline{f}$ is not flat at $\underline{x} = \underline{r}$, there is an integer K such that

$$\liminf_{i \to \infty} ||M^K(\underline{x}^i;\underline{r})\underline{f}(\underline{x}^i)|| > 0$$

for any sequence $\underline{x}^i \to \underline{r}$, $\underline{x}^i \in U_{\underline{r}}$. (The proof is given in [8, p. 337].)

From this theorem, we see that a deflation matrix with property (P) treats multiple zeros in a fashion similar to that of the deflation $1/(x-r)$ for functions of one variable. If we assume in addition that the deflation matrix is such that $M(\underline{x};\underline{r})$ and $M(\underline{x};\underline{s})$ commute for each $\underline{r},\underline{s} \in E^N$, then in any numerical calculations, we can deflate out multiple zeros using the same procedure as was outlined for the deflation $1/(x-r)$ in E^1. Clearly both norm deflation and inner product deflation have this commutative property.

A number of numerical results for deflation for nonlinear systems are reported by Brown and Gearhart [8, pp. 338-342]. In those results two basic iterative methods (the discretized Newton's method and Brown's method) were used in conjunction with three types of deflation: norm deflation with the ℓ_∞ norm, norm deflation with the Euclidean norm and gradient deflation. In order to have controlled numerical experiments the following strategy was used in all computations:

1) Given a nonlinear system and a starting guess $\underline{x}^0$, one of the solution techniques (discretized

Newton's or Brown's) was used to locate a first zero, $\underline{r}_1$.

2) One of the deflation techniques was then used in conjunction with the same method, <u>beginning with the same starting guess</u> $\underline{x}^0$ and an attempt was made to find a second zero.

3) If a second zero was found, deflation was performed again and an attempt was made to find a third zero. This process was continued, <u>always starting with the same</u> $\underline{x}^0$, until:

a) all zeros of the system (or the maximum number requested by the programmer) had been found;

b) the process diverged to "infinity";

c) the maximum number of iterations allowed was exceeded; or

d) the iterates reached a point at which the Jacobian matrix became singular.

To summarize a large number of numerical experiments, in no case was the same root found twice unless it happened to be a multiple root; that is, the deflation techniques presented here did indeed succeed in keeping iterates away from previously found roots. Of the three types of deflation studied, norm deflation using the ℓ_∞ norm was the most stable in addition to being the easiest to compute. We include one of the numerical examples here.

EXAMPLE [8, p. 341]. Consider the 3×3 system

$$f(x,y,z) = x^2 + 2y^2 - 4$$
$$g(x,y,z) = x^2 + y^2 + z - 8$$
$$h(x,y,z) = (x-1)^2 + (2y-\sqrt{2})^2 + (z-5)^2 - 4$$

which has two roots given by $\underline{r}_1 = (0,\sqrt{2},6)$ and $\underline{r}_2 = (2,0,4)$. This problem was run using seven values of $\underline{x}_0$ for Newton's Method and using two of the seven for Brown's method. The farthest starting guess from a solution for both methods was (1,1,1), which is about a distance of 5.1 (in Euclidean norm) from $\underline{r}_1$. The other principle starting guess used was (1,0.7,5), which is approximately midway between the two zeros (of Euclidean distance ≃1.58 from each). In all cases which were run, both zeros were found in at most 19 iterations per zero, except for Newton's Method starting at (1,1,1) used in conjunction with Euclidean norm deflation which failed to find the second zero in 200 iterations.

REFERENCES

[1] P. T. Boggs, The solution of nonlinear systems of equations by A-stable integration techniques, SIAM J. Numer. Anal., 8 (1971) 767-785.

[2] W. E. Bosarge, Jr., Infinite dimensional iterative methods and applications, Publication 320-2347, IBM Scientific Center, Houston, Texas, 1968.

[3] K. M. Brown, A quadratically convergent method for solving simultaneous nonlinear equations, Ph.D. Thesis, Purdue University, Lafayette, Ind., 1966.

[4] K. M. Brown, Solution of simultaneous nonlinear equations, Comm. ACM, 10 (1967) 728-729.

[5] K. M. Brown, A quadratically convergent Newton-like method based upon Gaussian elimination, SIAM J. Numer. Anal., 6 (1969) 560-569.

[6] K. M. Brown and S. D. Conte, The solution of simultaneous nonlinear equations, Proc. 22nd Nat. Conf. ACM, Thompson Book Co., Washington, D. C., 1967, 111-114.

[7] K. M. Brown and J. E. Dennis, Jr., On the second order convergence of Brown's derivative-free method for solving simultaneous nonlinear equations, Technical Report 71-7, Yale University Department of Computer Science, New Haven, Conn., 1971.

[8] K. M. Brown and W. B. Gearhart, Deflation techniques for the calculation of further solutions of a nonlinear system, Numer. Math., 16 (1971) 334-342.

[9] C. G. Broyden, A new method of solving simultaneous nonlinear equations, Computer J., 12 (1969) 94-99.

[10] P. Concus, Numerical solution of Plateau's problem, Math. Comp., 21 (1967) 340-350.

[11] S. D. Conte and C. deBoor, Elementary Numerical Analysis, Second Edition, McGraw-Hill, N. Y., 1972.

[12] G. G. Dahlquist, A special stability problem for linear multistep methods, BIT, 3 (1963) 27-43.

[13] D. F. Davidenko, On a new method of numerical solution of systems of nonlinear equations, Doklady Akad. Nauk SSSR (N.S.), 88 (1953) 601-602.

[14] J. Davis, The solution of nonlinear operator equations with critical points, Ph.D. Thesis, Oregon State University, Corvallis, 1966.

[15] J. E. Dennis, Jr., On the convergence of Newton-like methods, Numerical Methods for Nonlinear Algebraic Equations, P. Rabinowitz, ed., Gordon and Breach, London, 1970, 163-181.

[16] R. Fletcher, Function minimization without evaluating derivatives; a review, Computer J., 8 (1965) 33-41.

[17] G. E. Forsythe and C. B. Moler, Computer Solution of Linear Algebraic Systems, Prentice-Hall, Englewood Cliffs, N. J., 1967.

[18] F. B. Hildebrand, Introduction to Numerical Analysis, McGraw-Hill, N. Y., 1956.

[19] H. Kuhn, Approximate search for fixed points, Computing Methods in Optimization Problems II, A. Zadek, L. Neustadt and A. Balakrishnan, eds., Academic Press, N. Y., 1969, 199-211.

[20] G. H. Meyer, On solving nonlinear equations with a one-parameter operator imbedding, SIAM J. Numer. Anal., 5 (1968) 739-752.

[21] H. Möhlen, Lösung nichtlinearer Gleichungssysteme ein Vergleich des Hybridverfahrens von M.J.D. Powell mit dem Newton-ähnlich Verfahren von K. M. Brown, Diplomarbeit, Universität zu Köln, Germany, 1972.

[22] J. M. Ortega and W. C. Rheinboldt, Iterative Solution of Nonlinear Equations in Several Variables, Academic Press, N. Y., 1970.

[23] J. M. Ortega and M. L. Rockoff, Nonlinear difference equations and Gauss-Seidel type iterative methods, SIAM J. Numer. Anal., 3 (1966) 497-513.

[24] M. J. D. Powell, A FORTRAN subroutine for solving systems of non-linear algebraic equations, Report No. R-5947 A.E.R.E., Harwell, Didcot, Berkshire, England, 1968.

[25] M. J. D. Powell, A hybrid method for non-linear equations, Report No. T.P. 364, A.E.R.E., Harwell, Didcot, Berkshire, England, 1969.

[26] V. Samanskii, On a modification of the Newton method (Russian), Ukrain. Mat. Z., 19 (1967) 133-138.

[27] H. Scarf, The approximation of fixed points of a continuous mapping, SIAM J. Appl. Math., 15 (1967) 1328-1343.

[28] S. Schechter, Iteration methods for nonlinear problems, Trans. Amer. Math. Soc., 104 (1962) 179-189.

[29] M. A. Schumer and K. Steiglitz, Adaptive step size random search, IEEE Trans. Automatic Control, AC-13 (1968) 270-276.

[30] J. H. Wilkinson, Rounding Errors in Algebraic Processes, Prentice-Hall, Englewood Cliffs, N. J., 1963.

COMPUTER ALGORITHMS FOR SOLVING SYSTEMS OF NONLINEAR EQUATIONS

APPENDIX: FORTRAN IV PROGRAM FOR BROWN'S METHOD

```
C  SOLUTION OF SIMULTANEOUS NONLINEAR EQUATIONS WITHOUT REQUIRING DERIVATIVES
C                              BY
C                        KENNETH M. BROWN
C     DEPARTMENT OF COMPUTER, INFORMATION, AND CONTROL SCIENCES
C     INSTITUTE OF TECHNOLOGY       114    MAIN   ENGINEERING
C     UNIVERSITY OF MINNESOTA
C     MINNEAPOLIS, MINNESOTA  55455
C  SAMPLE CALLING PROGRAM
      DIMENSION X(4)
      X(1) = 0.
      X(2) = .01
      X(3) = 1.
      X(4) = .75
      CALL  NONLIN(4, 6,25,1,X,1.E-8)
      STOP
      END
C  END SAMPLE CALLING PROGRAM
C  IMPORTANT!  THE USER MUST FURNISH A SUBROUTINE NAMED AUXFCN
C              WHICH CONTAINS THE FUNCTIONS WHOSE ZEROS ARE SOUGHT.
C              AUXFCN SHOULD RETURN THE VALUE OF THE K-TH FUNCTION
C              OF THE SYSTEM EVALUATED AT THE VECTOR X. FOR A SAMPLE
C              OF THE WAY AUXFXN MUST BE WRITTEN, SEE THE LAST
C              SUBROUTINE IN THIS FILE.
C    NOTE:     FOR MAXIMUM EFFICIENCY ORDER YOUR FUNCTIONS IN
C              AUXFCN SO THAT THE LINEAR FUNCTIONS COME FIRST,
C              THEN THE FUNCTIONS BECOME PROGRESSIVELY MORE
C              NONLINEAR WITH THE "MOST NONLINEAR" FUNCTION COMING
C              LAST.
      SUBROUTINE NONLIN (N,NUMSIG,MAXIT,IPRINT,X,EPS)
C     THIS SUBROUTINE SOLVES A SYSTEM OF N SIMULTANEOUS NONLINEAR
C     EQUATIONS. THE METHOD USED IS AT LEAST QUADRATICALLY CONVERGENT AND
C     REQUIRES ONLY (N**2/2 + 3*N/2) FUNCTION EVALUATIONS PER ITERATIVE
C     STEP AS COMPARED WITH (N**2 + N) EVALUATIONS FOR NEWTON'S METHOD.
C     THIS RESULTS IN A SAVINGS OF COMPUTATIONAL EFFORT FOR SUFFICIENTLY
C     COMPLICATED FUNCTIONS.THE METHOD DOES NOT REQUIRE THE USER TO
C     FURNISH ANY DERIVATIVES.
C
C     REFERENCES:
C     1.  K. M. BROWN, SOLUTION OF SIMULTANEOUS NONLINEAR EQUATIONS,
C         COMM. OF THE ACM, VOL. 10, NO. 11, NOV., 1967, PP. 728-729.
C     2.  K. M. BROWN, A QUADRATICALLY CONVERGENT NEWTON-LIKE METHOD
C         BASED UPON GAUSSIAN ELIMINATION, SIAM J. ON NUMERICAL
C         ANALYSIS, VOL. 6, NO. 4, DECEMBER, 1969, PP. 560-569.
C
```

FORTRAN IV Program for Brown's Method

```
C       INPUT PARAMETERS FOLLOW.
C       N = NUMBER OF EQUATIONS ( = NUMBER OF UNKNOWNS).
C       NUMSIG = NUMBER OF SIGNIFICANT DIGITS DESIRED.
C       MAXIT = MAXIMUM NUMBER OF ITERATIONS TO BE ALLOWED.
C       IPRINT = OUTPUT OPTION, OUTPUT IF = 1; HOWEVER, FAILURE
C                INDICATIONS ARE ALWAYS OUTPUT : MAXIT EXCEEDED AND
C                SINGULAR JACOBIAN.
C       X = VECTOR OF INITIAL GUESSES.
C       EPS = CONVERGENCE CRITERION.  ITERATION WILL BE TERMINATED IF
C             ABS(F(I)) .LT. EPS, I=1,...,N,  WHERE F(I) DENOTES THE
C             I-TH FUNCTION IN THE SYSTEM.
C  NOTE!! CONVERGENCE CRITERION IS CONSIDERED TO BE MET IF EITHER
C         THE NUMBER OF SIGNIFICANT DIGITS REQUESTED IS ACHIEVED
C         OR THE EPS CRITERION ON THE FUNCTION VALUES IS SATISFIED.
C         TO FORCE THE ITERATION TO BE TERMINATED BY ONE OF THE
C         CRITERIA, SIMPLY SET THE OTHER ONE TO BE VERY STRINGENT.
C
C       OUTPUT PARAMETERS FOLLOW..
C       MAXIT = NUMBER OF ITERATIONS USED.
C       X = SOLUTION OF THE SYSTEM (OR BEST APPROXIMATION THERETO).
        REAL              X(30),PART(30),TEMP(30),COE(30,31),RELCON,F,
      1FACTOR,HOLD,H,FPLUS,DERMAX,TEST
        DIMENSION ISUB(30),LOOKUP(30,30)
C
C       FOR EXPOSITORY PURPOSES,COE AND LOOKUP ARE DIMENSIONED AT
C       30 X 31  AND  30 X 30  RESPECTIVELY.  CONSIDERABLE
C       STORAGE CAN BE SAVED AT THE EXPENSE OF MAKING THE
C       PROGRAM MORE DIFFICULT TO READ; IN FACT THE 930 LOCATIONS
C       FOR COE REDUCE TO 495 AND THE 900 LOCATIONS FOR LOOKUP
C       REDUCE TO JUST 30. A LISTING OF THIS CORE-WISE MORE
C       EFFICIENT VERSION OF THE ALGORITHM IS AVAILABLE FROM THE AUTHOR.
C
C  DELTA WILL BE A FUNCTION OF THE MACHINE AND THE PRECISION USED:
C
        DELTA=1.E-7
        RELCON=10.E+0**(-NUMSIG)
        JTEST = 1
        IF(IPRINT .EQ. 1) PRINT 48
 48     FORMAT (1H1)
        DO 700 M = 1, MAXIT
        IQUIT=0
        FMAX = 0.
        M1 = M-1
        IF (IPRINT .NE. 1) GO TO 9
        PRINT    49, M1, (X(I), I = 1,N)
 49     FORMAT(I5,3E18.8 / (E23.8 ,2E18. 8))
 9      DO 10 J = 1,N
 10     LOOKUP (1,J) = J
C
C       THE ARRAY LOOKUP PERMITS A PARTIAL PIVOTING EFFECT WITHOUT HAVING
C       TO PHYSICALLY INTERCHANGE ROWS OR COLUMNS.
C
        DO 500 K = 1,N
        IF (K-1) 134,134,131
```

```
 131    KMIN = K-1
        CALL BACK (KMIN,N,X,ISUB,COE,LOOKUP)
C
C       SET UP PARTIAL DERIVATIVES OF KTH FUNCTION..
C
 134    CALL AUXFCN (X,F,K)
        FMAX = AMAX1 (FMAX,ABS(F))
        IF (ABS(F) .GE. EPS) GO TO 1345
        IQUIT=IQUIT+1
        IF(IQUIT .NE. N) GO TO 1345
        GO TO 725
 1345   FACTOR=.001E+00
 135    ITALLY = 0
        DO 200 I = K,N
        ITEMP = LOOKUP (K,I)
        HOLD = X(ITEMP)
        PREC = 5.E-6
C       PREC IS A FUNCTION OF THE MACHINE SIGNIFICANCE, SIG, AND SHOULD BE
C       COMPUTED AS  PREC=5.*10.**(-SIG+2). IN THIS INSTANCE
C       WE WERE DEALING WITH AN 8 DIGIT MACHINE.
        ETA = FACTOR*ABS(HOLD)
        H = AMIN1 (FMAX,ETA)
        IF (H .LT. PREC) H=PREC
        X (ITEMP)=HOLD+H
        IF (K-1)161,161,151
 151    CALL BACK (KMIN,N,X,ISUB,COE,LOOKUP)
 161    CALL AUXFCN (X,FPLUS,K)
        PART (ITEMP) = (FPLUS-F)/H
        X (ITEMP) = HOLD
        IF ( ABS(PART(ITEMP)) .LT. DELTA  ) GO TO 190
        IF ( ABS(F/PART(ITEMP)) .LE. 1.E+15)GO TO 200
 190    ITALLY = ITALLY+1
 200    CONTINUE
        IF (ITALLY .LE. N-K) GO TO 202
        FACTOR = FACTOR*10.0E+00
        IF (FACTOR .GT. 11.) GO TO 775
        GO TO 135
 202    IF (K .LT. N) GO TO 203
        IF ( ABS(PART(ITEMP)).LT. DELTA ) GO TO 775
        COE (K,N+1) = 0.0E+00
        KMAX = ITEMP
        GO TO 500
C
C       FIND PARTIAL DERIVATIVE OF LARGEST ABSOLUTE VALUE..
C
 203    KMAX = LOOKUP (K,K)
        DERMAX = ABS (PART(KMAX))
        KPLUS = K+1
        DO 210 I = KPLUS,N
        JSUB = LOOKUP (K,I)
        TEST = ABS (PART(JSUB))
        IF (TEST .LT. DERMAX) GO TO 209
        DERMAX = TEST
        LOOKUP (KPLUS,I) = KMAX
        KMAX = JSUB
        GO TO 210
```

FORTRAN IV Program for Brown's Method

```
 209    LOOKUP (KPLUS,I) = JSUB
 210    CONTINUE
        IF ( ABS(PART(KMAX)) .EQ. 0.0) GO TO 775
C
C       SET UP COEFFICIENTS FOR KTH ROW OF TRIANGULAR LINEAR SYSTEM USED
C       TO BACK-SOLVE FOR THE FIRST K  VALUES OF X(I)
C
        ISUB (K) = KMAX
        COE (K,N+1) = 0.0E+00
        DO 220 J = KPLUS,N
        JSUB = LOOKUP (KPLUS,J)
        COE (K,JSUB) = -PART (JSUB)/PART(KMAX)
        COE (K,N+1) = COE (K,N+1)+PART (JSUB)*X(JSUB)
 220    CONTINUE
 500    COE (K,N+1) = (COE (K,N+1)-F)/PART(KMAX)+X(KMAX)
C       BACK SUBSTITUTE TO OBTAIN NEXT APPROXIMATION TO X:
        X (KMAX) = COE (N,N+1)
        IF (N .EQ. 1) GO TO 610
        CALL BACK (N-1,N,X,ISUB,COE,LOOKUP)
 610    IF (M-1) 650,650,625
C
C       TEST FOR CONVERGENCE..
C
 625    DO 630 I = 1,N
        IF( ABS(TEMP(I)-X(I)) .GT. ABS(X(I))*RELCON ) GO TO 649
 630    CONTINUE
        JTEST = JTEST+1
        IF (JTEST-3)650,725,725
 649    JTEST = 1
 650    DO 660 I = 1,N
 660    TEMP (I) = X(I)
 700    CONTINUE
        PRINT 1753
 1753   FORMAT(/' NO CONVERGENCE. MAXIMUM NUMBER OF ITERATIONS USED.')
        IF(IPRINT .NE. 1) GO TO 800
        PRINT 1763
 1763   FORMAT('FUNCTION VALUES AT THE LAST APPROXIMATION FOLLOW:'/)
        IFLAG=1
        GO TO 7777
 725    IF (IPRINT .NE. 1) GO TO 800
 7777   DO 750 K = 1,N
        CALL AUXFCN (X,PART(K),K)
 750    CONTINUE
        IF(IFLAG .NE. 1) GO TO 8777
        PRINT 7788,(PART(K),K=1,N)
 7788   FORMAT(3E20.8)
        GO TO 800
 8777   PRINT 751
  751   FORMAT(//' CONVERGENCE HAS BEEN ACHIEVED. THE FUNCTION VALUES')
        PRINT 7515,(PART(K),K = 1,N)
 7515   FORMAT(' AT THE FINAL APPROXIMATION FOLLOW:'//(3E20.8))
        GO TO 800
```

COMPUTER ALGORITHMS FOR SOLVING SYSTEMS OF NONLINEAR EQUATIONS

FORTRAN IV Program for Brown's Method

```
 775    PRINT  752
 752    FORMAT(//'MODIFIED JACOBIAN IS SINGULAR.  TRY A DIFFERENT')
        PRINT 7525
 7525   FORMAT('INITIAL APPROXIMATION.')
 800    MAXIT=M1 + 1
        RETURN
        END

        SUBROUTINE BACK (KMIN,N,X,ISUB,COE,LOOKUP)
C       THIS SUBROUTINE BACK-SOLVES THE FIRST KMIN ROWS OF A TRIANGULARIZED
C       LINEAR SYSTEM FOR IMPROVED X VALUES IN TERMS PREVIOUS ONES.
        DIMENSION        X(30),COE(30,31)
        DIMENSION ISUB(30),LOOKUP(30,30)
        DO 200 KK = 1,KMIN
        KM = KMIN-KK+2
        KMAX = ISUB (KM-1)
        X (KMAX) = 0.0E+00
        DO 100 J = KM,N
        JSUB = LOOKUP (KM,J)
        X (KMAX) = X (KMAX)+COE(KM-1,JSUB)*X(JSUB)
 100    CONTINUE
        X(KMAX)=X(KMAX)+COE(KM-1,N+1)
 200    CONTINUE
        RETURN
        END
C  SAMPLE AUXFCN FOLLOWS
        SUBROUTINE AUXFCN(X,Y,K)
        DIMENSION        X(4)
        T=X(4)
        GO TO (1 ,2 ,3 ,4),K
 1      Y=(T/3.)*X(1) + (T/6.)*X(2)-T**3/12.
        RETURN
 2      Y=(T/6.)*X(1)+X(2)/3.+(1.-T)*X(3)/6.
      1 -(T**2+T+1.)/12.
        RETURN
 3      Y=(1.-T)*X(2)/6. + (1.-T)*X(3)/3.+(T**3+T**2+T-3.)/12.
        RETURN
        Y=3.*(X(3)-X(1) )*T**2+2.* (X(3)-X(2))*T+X(3)-X(2)+2.
      1 *(X(1)**2 -X(3)**2)+2.*X(2)*(X(1)-X(3))
        RETURN
        END
```

COMPUTER ALGORITHMS FOR SOLVING SYSTEMS OF NONLINEAR EQUATIONS

FORTRAN IV Program for Brown's Method

Output From Sample Problem

ITER.#	X(1) X(4)	X(2)	X(3)
0	0. 7.50000000E-01	1.00000000E-02	1.00000000E+00
1	-1.29794926E-01 2.75670858E-01	1.30101573E-02	8.98745049E-01
2	5.85514940E-02 4.16498247E-01	1.05151556E-01	9.43530710E-01
3	-1.23766444E-01 6.54815498E-01	3.56132871E-01	9.93628190E-01
4	-5.90355479E-02 5.10017779E-01	2.14111251E-01	9.58542460E-01
5	-4.21861600E-02 5.00544750E-01	2.08960860E-01	9.58431847E-01
6	-4.16677032E-02 5.00002348E-01	2.08335646E-01	9.58333905E-01
7	-4.16666667E-02 5.00000000E-01	2.08333333E-01	9.58333333E-01

CONVERGENCE HAS BEEN ACHIEVED. THE FUNCTION VALUES
AT THE FINAL APPROXIMATION FOLLOW.

5.55111512E-17 1.77635684E-15 8.99178420E-16
3.50475204E-12

REMARK: THE SAMPLE PROBLEM SOLVED HERE WAS THE PROBLEM OF DETERMINING THE OPTIMAL KNOT LOCATION AND COEFFICIENTS WHEN FITTING THE FUNCTION Y = X**2 BY LINEAR SPLINES IN THE LEAST SQUARES SENSE ALLOWING ONE INTERNAL VARIABLE KNOT LOCATION. AS THE OUTPUT SHOWS, THE KNOT SHOULD BE LOCATED AT .5 AND THE COEFFICIENTS OF THE BASIS FUNCTIONS SHOULD BE -.041666667, .20833333 AND .95833333.

ON THE CHOICE OF RELAXATION PARAMETERS FOR NONLINEAR PROBLEMS

Samuel Schechter*

1. Introduction.

To obtain global convergence for relaxation methods the relaxation parameters must generally be chosen from an interval dictated by the iterate as well as the smoothness of the functions involved. In [4], [6] it was shown that for sufficiently smooth functions, constructive techniques may be given to determine this interval for each iterate. Our objective here will be to indicate an extension of these processes to those not requiring second derivative computation at each step. These methods may be regarded as modified relaxation methods which

---

*This work was supported in part by the U. S. Army Research Office - Durham, under contract DAHC04-72-C-0030.

include SOR-one-step secant (or Steffenson). Proofs of global convergence will be indicated for these methods.

A theorem is obtained for determining bounded level sets and the methods are illustrated by several examples taken from the variational calculus.

2. The Problem.

We assume that a real valued function $g(u)$ is given, to be minimized over an open set, which, for simplicity, we take to be the whole space R^n. We assume $g \in C^1(R^n)$ and denote its gradient vector by $r(u) = \text{grad } g(u)$. For a given $u^o \in R^n$ we define the level set

$$D(g;u^o) = \{u: g(u) \leq g(u^o)\} \tag{2.1}$$

(which we will also denote by $D(u^o)$ or simply D). We thus seek a minimum in D and thus a global solution u^* to the system

$$r(u^*) = 0 \tag{2.2}$$

For this vector u^o and for some chosen coordinate unit vector e^i, let δ_o and ω_o be given positive numbers. We may then move in the direction of this unit vector to get a new guess

$$u^1 = u^o + \omega_o d_o e^i$$

where $d_o = -r_i(u^o)/\delta_o$. We will call this a modified relaxation step if $0 < \omega_o < 2$. We wish to choose ω_o so that $g(u^1) \leq g(u^o)$. To achieve this we make some further assumptions about g.

Let $I_o = [u^o, u^o + 2d_o e^i]$ denote the closed line segment joining the two points in the brackets. We assume that $a_i(u) = \partial_i r_i(u)$ exists and is positive in D and bounded on I_o. We use the notation $||f||_B = \sup_B |f(x)|$ for any function on a set B with values in R^n.

We let

$$(2.3)\quad \begin{cases} \Lambda_o = \max(\delta_o, ||a_i||_{I_o}) \\ \\ \gamma_o = \delta_o/\Lambda_o \end{cases}$$

Then as in [6], we may achieve our goal if ω_o is chosen in the interval

$$(2.4) \qquad 0 < \omega_o < 2\gamma_o \leq 2.$$

For by expanding g about u^o along I_o as a function of u_i, we get for some $v \in I_o$,

$$\begin{aligned}
-\Delta g_o &= g(u^o)-g(u^1) \\
&= r_i(u^o)(u^1-u^o)^T e^i + \frac{1}{2} a_i(v)|u^1-u^o|^2 \\
&= (r_i(u^o))^2 \omega_o(2-\omega_o a_i(v)/\delta_o)/2\delta_o \\
&= (\omega_o/2\gamma_o\Lambda_o)(2-\omega_o a_i(v)/\gamma_o\Lambda_o)(r_i(u^o))^2 \\
&\geq (\hat{\omega}_o/2\Lambda_o)(2-\hat{\omega}_o)(r_i(u^o))^2 \geq 0
\end{aligned}$$

where $\hat{\omega}_o = \omega_o/\gamma_o$. Thus $g(u^o) \geq g(u^1)$ and equality holds only when $r_i(u^o) = 0$.

3. The Parameter δ_o.

The usual choice for δ_o is $a_i(u^o)$ as in [4], [6]. We will now consider other methods of choosing

δ_o. Assume that we have two guesses u^o and $\tilde{u}^o$ such that $g(u^o) \leq g(\tilde{u}^o)$, and such that $u^o - \tilde{u}^o$ has only its ith component nonzero. If this is not so the roles of the two guesses are interchanged. If we wish to avoid this test entirely we may use a preliminary modified relaxation step starting with $\tilde{u}^o$ and some $\tilde{\delta}_o$ to get

$$(3.1) \qquad u^o = \tilde{u}^o - \tilde{\omega}_o r_i(\tilde{u}^o) e^i / \tilde{\delta}_o$$

for some $r_i(\tilde{u}_o) \neq 0$. (To avoid computing Λ_o we may use an available uniform upper bound on $a_i(u)$. We may then choose $\tilde{\delta}_o$ larger than this bound, so that $\Lambda_o = \tilde{\delta}_o$.)

A new guess is generated by another modified relaxation step using difference quotients:

$$\Delta r_i{}^o = r_i(u^o) - r_i(\tilde{u}^o), \quad \Delta u_i{}^o = u_i{}^o - \tilde{u}_i{}^o$$

$$(3.2) \qquad \delta_o = \begin{cases} \delta_{01} = (\Delta r_i{}^o)/(\Delta u_i{}^o), & |u^o - \tilde{u}^o| > \varepsilon \\ \delta_{02} = a_i(u^o), & |u^o - \tilde{u}^o| \leq \varepsilon \end{cases}$$

where $\varepsilon \geq 0$ is some given tolerance. It is clear that $\delta_o > 0$ since we assume $a_i(u) > 0$ in $D(\tilde{u}^o)$. Thus we get a new iterate

$$(3.3) \qquad \tilde{u}^1 = u^o - \omega_o r_i(u^o)e^i/\delta_o$$

such that $g(\tilde{u}^1) \leq g(u^o)$. We then repeat the previous step with a new direction e^j, and a newly generated u^1. This yields a sequence $\{u^p\}$ such that $g(u^{p+1}) \leq g(u^p)$, if (2.3) and (2.4) are used.

If $\varepsilon = 0$ then we obtain what may be called a relaxed secant method. If (3.1) is used at each step with $\tilde{\delta}_p = a_i(u^p)$ we get a relaxed Steffenson method. Note also that the process may proceed without actual evaluations of $a_i(u)$, unless the tolerance is too large.

In using (3.3) it is understood that the previous interval I_o is replaced by one using the δ_o given by (3.2).

4. A Convergence Result.

If we impose more stringent conditions on g

we obtain a more flexible range for ω_o. Let us assume as in [4] that $\partial_i a_i(u) = b_i(u)$ exists in $D(\tilde{u}^o)$ and that $|b_i(u)| \leq \mu_o$ in I_o. By using the Taylor expansion again, but using the cubic term we get from (3.3)

$$(4.1) \quad -\Delta g_o = g(u^o) - g(\tilde{u}^1) = \frac{r_i(u^o)^2 \omega_o}{2\delta_o}\left(2 - \frac{a_i(u^o)}{\delta_o}\omega_o - \omega_o^2 \frac{b_i(v)d_o}{3\delta_o}\right)$$

for v in I_o.

If $\delta_o = a_i(u^o)$ then we choose

$$(4.2) \quad \gamma_o = \frac{2}{1 + \sqrt{1 + \frac{8}{3}\psi_o}}$$

where

$$(4.3) \quad \psi_o = \mu_o |d_o| / \delta_o .$$

If $\delta_o = \delta_{01}$ then we choose

$$(4.4) \quad \gamma_o = \frac{2}{\alpha_o + \sqrt{\alpha_o^2 + 8\psi_o/3}}, \text{ if } \alpha_o \geq 1$$

where $\alpha_o = a_i(u^o)/\delta_o$. The same choice of γ_o may be used for any other modified relaxation. In either case, if

$$0 < \omega_o < 2\gamma_o$$

then, $g(\tilde{u}^1) \leq g(u^o)$ since

$$-\Delta g_o \geq \frac{\omega_o r_i(u^o)^2}{2\delta_o} (2 - \alpha_o \omega_o - \omega_o^2 \psi_o/3)$$

$$\alpha_o + \omega_o \psi_o/3 < \alpha_o + \frac{2\gamma_o \psi_o}{3} = \frac{1}{\gamma_o}$$

we get

$$2 - \omega_o(\alpha_o + \omega_o \psi_o/3) \geq 2 - \hat{\omega}_o$$

and

$$-\Delta g_o \geq \frac{\omega_o r_i(u^o)^2}{2\delta_o} (2 - \hat{\omega}_o) \geq 0.$$

In order to avoid computing α_o or $a_i(u^o)$ in (4.4) we may replace α_o by some upper bound of this quotient, and get the same result.

If $\alpha_o < 1$ then we may choose γ_o from (4.2) since then $2 - \alpha_o > 2 - \omega_o$. In either case we get $\gamma_o \leq 1$.

If we happen to get $b_i(v)r_i(u^o) \geq 0$ then we may choose $\gamma_o = \min(1, \frac{1}{\alpha_o})$. This follows directly from (4.1). In particular, if we use the usual relaxation where $\alpha_o = 1$, we get $\gamma_o = 1$ and thus the full possible range for ω_o.

These results extend those given in [4]. It is worth noting that as $r_i(u^p) \to 0$ we get $\psi_p \to 0$ and $\alpha_p \to 1$. This means that the $\gamma_p \to 1$ or that the range of ω_p expands to the interval (0,2) of the linear case.

For many problems it has been shown [3] that the asymptotic rate of convergence, for cyclic ordering, is governed by the optimal parameter ω_b* determined by the Hessian matrix r'(u*) and a corresponding linearized problem at the solution. For linear elliptic problems this value is usually larger than one: $\omega_b^* = 1+\delta$, $1 > \delta > 0$. Unless γ_p is large enough, this parameter value is not usable. If we are using, say, (4.2) then this value will be usable when $\psi_p < 3\left[\frac{1-\delta}{(1+\delta)^2}\right]$. This is easily checked as the iteration progresses.

We note also that $g(u)$ need not be convex in order for the iterates to descend. We have used only that $g(u)$ is convex in each variable separately.

However, in order to assure convergence of the iterates, further assumptions must be made:

THEOREM 4.1. If the initial level set D is bounded and, if a cyclic ordering is used in selecting the e^i, then the ω_p may be chosen so that $r(u^p) \to 0$. If, furthermore, it is assumed that $g \in C^2(D)$ and that the solution to (2.2) is unique then the u^p will converge to the solution u^*.

The proofs are essentially the same as those given in [4] and [6] and we refer the reader to those sources.

A scheme related to the Steffenson format was suggested by Kronrod [2], for solving nonlinear elliptic equations. It did not involve any relaxation parameters nor were any proofs given.

5. Group Relaxation.

For group relaxation we consider only the usual format of the iteration, where the elements of the Hessian are computed. That is, we assume a multiindex k is given and, using the notation of [6], we assume that the principal submatrix $A_k(u^o)$ of the Hessian of g(u) exists and is positive definite. We may then define for a given guess u^o, and ω_o

$$(5.1)\qquad \begin{cases} u_k^1 = u_k^o + \omega_o d_o \\ \\ u_{k'}^1 = u_{k'}^o \end{cases}$$

where $d_o = -A_k^{-1}(u^o)\cdot r_k(u^o)$ and where k' is the complementary multiindex to k, while r_k is the corresponding subvector. We let v^o be the vector obtained from the right side of (5.1) when ω_o is replaced by 2. Let $I_o = [u^o, v^o]$ and assume that $A_k(u)$ is positive definite $A_k > 0$ on I_o. We further assume that for $j \varepsilon k$, the matrices $\partial_j A_k(u) = B_{j,k}(u)$ exist on

I_o and have there a bound

$$\sum_{j \varepsilon k} |B_{j,k}(u)|^2 \leq \mu_o^{\,2}$$

where the spectral norm of the matrix is used for the absolute value.

If we let $\phi_o(v) = (d_o, A_k(v)d_o)/(d_o, d_o)$ for any v in I_o then we define as in the previous section

$$\psi_o = \mu_o |d_o| / \phi_o(u^o).$$

We can then show

THEOREM 5.1. If

$$\gamma_o = \frac{2}{1 + \sqrt{1 + \frac{8}{3}\psi_o}}$$

and ω_o is chosen in the interval

$$0 < \omega_o < 2\gamma_o$$

then

$$g(u^1) \leq g(u^o),$$

with equality, only if $d_o = 0$, or $r_k(u^o) = 0$.

PROOF: We use Taylor's expansion about u^o along I_o:

$$g(u^1)-g(u^o) = (g'(u^o),u^1-u^o) + \frac{1}{2}(u^1-u^o)^T g''(u^o)(u^1-u^o)$$

$$+ \frac{1}{6} \sum_{j\varepsilon k} (u^1-u^o)^T B_j(z)(u^1-u^o)\ (u^1-u^o)_j$$

for $z \varepsilon I_o$ or that

$$-\Delta g_o = \frac{1}{2}\omega_o(d_o,d_o)\bullet$$

$$\left(2\phi_o(u^o)-\omega_o\phi_o(u^o) - \frac{\omega_o^2}{3} \sum_{j\varepsilon k} \frac{(d_o,B_j(z)d_o)\cdot(d_o)_j}{(d_o,d_o)}\right)$$

$$\geq \frac{1}{2}\ \omega_o(d_o,d_o)\phi_o(u^o)[2 - \omega_o - \frac{\omega_o^2}{3}\ \psi_o]$$

$$\geq \frac{1}{2}\ \omega_o(d_o,A_k(u^o)d_o)[2 - \hat{\omega}_o]$$

which gives the result.

The remarks at the end of the previous section apply equally to this group relaxation case. In both cases other orderings may be used depending on g, [5], [6].

6. Applications

We will examine several applications of the previous results. As a first example we consider the semilinear elliptic equation $-\Delta\phi + f(\phi) = 0$ in two dimensions with Dirichlet data. If we approximate the problem in the simplest way by a five point difference scheme [5], we get for mesh width h, that $a_i(u) = 4 + h^2 f'(u_i) \geq 4$, if $f'(\phi) \geq 0$. We set $u_i = \phi(P_i)$, for P_i a mesh point. Let us assume that $f''(\phi) \geq 0$ for all ϕ, so that $b_i(u) = h^2 f''(u_i) \geq 0$.

If for some u^o we get $r_i(u^o) \geq 0$, then $b_i(v) r_i(u^o) \geq 0$ and we may set $\gamma_o = 1$. If $r_i(u^o) < 0$ then $d_o > 0$ and $u_i^o < u_i^o + 2d_o$. If we use (2.3) then since $a_i(u)$ is increasing on I_o

$$\Lambda_o = \max(\delta_o, \ a_i(u^o + 2d_o e^i)).$$

If $\delta_o = a_i(u^o)$ then

$$\gamma_o = \frac{a_i(u^o)}{a_i(u^o + 2d_o e^i)} .$$

similar situation is found if $f'(\phi)$ is decreasing.

If we assume that $f''(\phi)$ is monotone increasing then we may choose μ_o to get γ_o from (4.2), if $\alpha_o = 1$. Let $M_o = \max(u_i^o, u_i^o + 2d_o)$ then $\mu_o = b_i(M_o)$.

To be more specific, let $f(\phi) = e^{\phi}$ then the previous criteria are satisfied.

If we take $\alpha_o = 1$ then we get as above:

$$\gamma_o = 2\left\{1 + \left[1 + \frac{8}{3}\,\frac{h^2 e^{M_o}|d_o|}{4 + h^2 e^{u_i^o}}\right]^{1/2}\right\}^{-1}.$$

Since we need only use this when $r_i(u^o) < 0$ or when $d_o > 0$, this gives us $M_o = u_i^o + 2d_o$. Thus we see that for $f(\phi) = e^{\phi}$, the range of ω_o can be easily computed by checking the sign of $r_i(u^o)$.

A more general example is obtained by discretization of a variational problem of the form $\int F(p,q)\,dxdy = \min.$, for ϕ given on the boundary of the region of integration and $\phi_x = p$, $\phi_y = q$. When the simplest difference method is used, we get a finite sum, as in [5]

$$g(u) = h^2 \Sigma_i F(p_i, q_i) \tag{6.1}$$

where

$$\begin{cases} p_i = (u_{i1} - u_i)/h \\ q_i = (u_{i2} - u_i)/h \end{cases} \tag{6.2}$$

and where $u_{i\alpha} = \phi(P_{i\alpha})$, $P_{i1}, \ldots, P_{i4}$ being the four neighbors of $P_{i0} = P_i$ on the mesh.

If we assume $F \varepsilon C^2(R^2)$ in its p,q variables then we denote its Hessian matrix by

$$\Phi = \begin{bmatrix} F_{pp} & F_{pq} \\ F_{pq} & F_{qq} \end{bmatrix} = \Phi(p,q)$$

A simple computation gives as in [5]

$$a_i(u) = \tag{6.3}$$

$$2\{e^T \Phi(p_i, q_i) e + F_{pp}(p_{i3}, q_{i3}) + F_{qq}(p_{i4}, q_{i4})\}$$

where $e = (1,1)^T$, so that $a_i(u) > 0$ for all u if

$\Phi > 0$ for all p,q. It is apparent from (6.3) that it is sufficient if the main diagonal terms of Φ and the sum of its entries are positive for all p,q.

We are also interested in upper bounds on a_i, b_i and determining when level sets are bounded.

7. Criterion for Bounded Level Sets.

To check if a level set is bounded we may use the following:

LEMMA 7.1. Let T be a transformation $T: K \supset R^n \to M \in R^m$ such that for all $u,v \in K$

$$(7.1) \qquad |T(u)-T(v)| \geq \delta(|u-v|)$$

where $\delta(\varepsilon): [0,\infty] \to [0,\infty]$ is a strictly increasing nonnegative (forcing) function such that $\delta(0) = 0$, $\delta(\infty) = \infty$. Let $h: M^o \to R^1$ be a continuous functional such that $M^o \supset M$ and such that for all $v^o \in M^o$, $M_1 = M^o \cap D(h;v^o)$ is bounded. Let $g(u) = h(T(u))$ then $K \cap D(g;u^o) = K_1$ is bounded for all u^o in K.

PROOF: Let $v^o = T(u^o)$ and let, for any u in

K_1, $v = T(u)$. Thus $v, v^o \varepsilon M$ and since

$$g(u) = h(T(u)) \leq h(v^o) = g(u^o)$$

we get that $v \varepsilon M_1$. Since M_1 is bounded, there is a positive number α such that for all $v \varepsilon M_1$

$$\alpha \geq |v-v^o| = |T(u)-T(u^o)| \geq \delta(|u-u^o|)$$

or $\delta^{-1}(\alpha) \geq |u-u^o|$.

Since u was arbitrary the lemma is established.

To apply this, we set

$$H(v) = h^2 \Sigma_i F(p_i, q_i)$$

where we set $v_i = (p_i, q_i)$, $v \varepsilon R^N$. We may from (6.2) define, for some matrix Q

$$v = Qu+c = Tu$$

for some constant c. Since $|Qu|$ is a quadratic Dirichlet norm, there is a positive number λ such that

$$|Qu| \geq \lambda |u|.$$

Thus T is an affine transformation satisfying (7.1) with $\delta(\varepsilon) = \lambda\varepsilon$, for all $u \in R^n$.

If we can show that all the level sets of $H(v)$ are bounded we will get the same for $g(u)$. A sufficient condition for $H(v)$ to have this property is that $\Phi \geq \mu > 0$ for a constant μ for all p,q. This follows, as in [5], from the fact that $H(v)$ is then a uniformly convex function in v.

A weaker condition would be to assume that $F(p,q) \to \infty$ when $|(p,q)| \to \infty$. We will call a real function $f(z)$, $z \in R^n$ <u>upward mobile</u> (or $\varepsilon\ (UM)_n$) if $f(z) \to \infty$ when $|z| \to \infty$. The bounded level sets of h will follow from the following:

LEMMA 7.2. Let $f(v) \in (UM)_n$, $g(w) \in (UM)_m$ and assume f and g are continuous. Assume further that $D_1 = D(f;v^o)$ and $D_2 = D(g;w^o)$ are bounded level sets for every $v^o \in R^n$, $w^o \in R^m$. Let

$$h(v,w) = f(v) + g(w)$$

then $D_3 = D(h;(u^o,w^o))$ are bounded for all (u^o,w^o).

PROOF: We let $g_* = \inf_{D_2} g(w)$ for a fixed v^o, w^o. Thus for any (v,w) in D_3, v is in the set

$$\{v: f(v) \leq f(v^o) + g(w^o) - g_*\}.$$

Since f is UM there is a v' such that $f(v') = f(v^o) + g(w^o) - g_* \geq f(v^o)$; $D(f;v')$ is bounded and contains v. Similarly there is a $D(g;w')$ which is bounded and contains any w component of D_3. Thus

$$D_3 \subset D(g;w') \times D(f;v')$$

which is bounded.

A similar proof gives the result for any finite sum of functions.

A simple example for $m = n = 1$ shows that without the UM condition, the lemma is false. Take $f(v) = v^2/(1+v^2)$ and $g(w) = f(w)$. For $(v^o,w^o) = (1,1)$, $f(v) + f(w) \leq 1 = 2f(1)$ is not bounded while $f(v) \leq f(1) = \frac{1}{2}$ is bounded. (For functions of the same variable it is not necessary to have the UM condition.)

THEOREM 7.1. If $F(p,q)$ is (UM) and continuous in (p,q) such that $D(F;(p^o,q^o))$ is bounded for every p^o,q^o then $D(g;u^o)$ is bounded for every u^o.

This follows readily from the previous lemmas. We note also that the latter theorem applies if q_i is replaced by u_i, and p_i may be a vector. Thus the result yields bounded level sets for higher dimensional variational problems as well as for higher order elliptic equations.

Other results in this direction have been obtained by Stepleman [7].

8. Examples.

A common special case of (6.1) is when $F(p,q) = f(\omega)$, $\omega = p^2 + q^2$. For the minimal surface equation [5], $f = (1+\omega)^{1/2}$. In magnetostatics [1] we find a function f whose derivative is $f' = (\varepsilon+\omega)/(1+\omega)$, $1 >> \varepsilon > 0$.

A simple calculation gives for $f_{i\alpha} = f(\omega_{i\alpha})$ and $f' = D_\omega f$:

$$r_i(u) = 2h[q_{i4}f'_{i4} + p_{i3}f'_{i3} - (p_i+q_i)f'_i]$$

$$a_i(u) = 2[2p_{i3}^2 f''_{i3} + f'_{i3} + 2q_{i4}^2 f''_{i4} + f'_{i4}$$

$$+ 2f'_i + 2(p_i+q_i)^2 f''_i]$$

$$b_i(u) = \frac{4}{h}[3p_{i3}f''_{i3}+2p_{i3}^3 f'''_{i3}+3q_{i4}f''_{i4}+2q_{i4}^3 f'''_{i4}$$

$$- 6(p_i+q_i)f''_i - 2(p_i+q_i)^3 f'''_i].$$

Similarly the matrix Φ takes a simple form

$$\Phi = 2(2f''(\omega)zz^T + f'(\omega)\cdot I_2)$$

where $z = (p,q)^T$ and I_2 is the unit matrix of order two. Since $zz^T \geq 0$ we will get $\Phi \geq \lambda > 0$ if, say, $f'' \geq 0$ and $f' \geq \lambda > 0$.

For $f = (1+\omega)^{1/2}$ we get that for all u, $0 < a_i(u) \leq 4$ since $f'' < 0$, and $|b_i(u)| \leq \frac{3}{h}(1 + \sqrt{2}) = \mu_o$. Since (r_i/h) is bounded for all u_o, $|\mu\cdot r_i|$ and thus ψ_o is bounded. Thus unless a_i is small, ψ_o will go to zero with r_i. One can expect therefore to take small steps when ω is large since then a_i is small. Otherwise $\gamma_p \to 1$ with

$r_i \to 0$.

For the function $f'(\omega) = (\varepsilon+\omega)/(1+\omega)$ we get that $a_i(u) \leq 28$ for all u, $|b_i(u)| \leq 4(7\sqrt{2}+10)(1-\varepsilon)/h = \mu_o$.

It is of interest to note that for the minimal surface $f(\omega)$ is UM and has bounded level sets. Thus from our theorem, the composite function $g(u)$ has only bounded level sets. A similar situation applies to the second example since $f = \omega - (1-\varepsilon)\log(1+\omega)$ which is UM.

REFERENCES

1. P. Concus, Numerical solution of the nonlinear magnetostatic-field equation in two dimensions, J. Comput. Phys. 1 (1967) 330-342.

2. A. S. Kronrod, Numerical solution to the equation of the magnetic field in iron with allowance for saturation, Soviet Phys. Dokl. 5 (1960) 513-514.

3. J. M. Ortega and W. C. Rheinboldt, Iterative Solution of Nonlinear Equations in Several Variables, Academic Press, N.Y., 1970.

4. S. Schechter, Minimization of a convex function by relaxation, Chap. 7 in Integer and Nonlinear Programming, J. Abadie ed., North Holland, Amsterdam 1970 pp. 177-189.

5. S. Schechter, Iteration methods for nonlinear problems, Trans. Amer. Math. Soc. 104 (1962) 179-189.

6. S. Schechter, Relaxation methods for convex problems, SIAM J. Numer. Anal. 5 (1968) 601-612.

7. R. S. Stepleman, Finite dimensional analogues of variational problems in the plane, SIAM J. Numer. Anal. 8 (1971) 11-23.

THE CONTRACTOR THEORY OF SOLVING EQUATIONS

Mieczyslaw Altman

0. Introduction.

The concept of contractors was introduced [6] as a tool for solving general equations in Banach spaces. Thus, various existence theorems for the solution of equations can be obtained along with convergence theorems for a broad class of iterative procedures, which differ in nature. The notions of contractors and nonlinear majorant functions can be easily combined to give both existence and convergence theorems. These results, which reveal the character of the convergence, are the basis for a unified theory for a large class of iterative methods, including the most important ones. The above topics are treated in §1, along with inverse derivatives.

In §2, directional contractors [5] are considered as a generalization of contractors and are used to obtain a generalization of the Banach fixed point theorem. An application, in the form of an existence theorem for nonlinear evolution equations, is also considered.

Finally, §3 is concerned with contractors which have nonlinear majorant functions.

1. Inverse Derivatives and Contractors [6].

1.1. Inverse derivatives.

Let $P: X \to Y$ be a nonlinear operator from a Banach space X to a Banach space Y. Consider the difference $P(x+h)-Px = Q(x)h$ and suppose that $\Gamma(x)$ is a linear bounded operator associated with $x \in X$ acting from Y to X, i.e. $\Gamma(x): Y \to X$.

DEFINITION 1.1.1. If $\Gamma(x)$ has the property: for $y \in Y$

$$(1.1.1) \qquad ||y||^{-1}||Q(x)\Gamma(x)y - y|| \to 0 \text{ as } y \to 0,$$

then $\Gamma(x)$ is called an <u>inverse derivative</u> at x of P. Condition (1.1.1) can be written in the form

(1.1.2) $$||y||^{-1}||P(x+\Gamma(x)y)-Px-y|| \to 0 \text{ as } y \to 0.$$

EXAMPLE. Let F be a nonlinear functional defined on X, $F: X \to R$ (reals). If the gradient $F'(x)$ exists and $h \in X$ is such that $F'(x)h \neq 0$, then we have

$$t^{-1}|F(x-t[F'(x)h]^{-1}h)-F(x)-t| \to 0 \text{ as } t \to 0,$$

where $t \in R$, i.e., according to (1.1.2),

$$\Gamma(x)t = t(F'(x)h)^{-1}h: R \to X$$

is an inverse derivative of F at x, if h is fixed and $F'(x)h \neq 0$.

Inverse derivatives have the following properties.

(i) If $\Gamma(x)$ exists, then $\frac{d}{dt}P(x+t\Gamma(x)y)|_{t=0} = y$, i.e. P has a directional derivative in the direction $\Gamma(x)y$.

(ii) Property (i) shows that $\Gamma(x)y = 0$ implies $y = 0$, i.e. $\Gamma(x)$ is a one-to-one mapping and $\Gamma(x)^{-1}$ is continuous.

(iii) If the Fréchet derivative $P'(x)$ exists, then

$P'(x)$ is an extension of $[\Gamma(x)]^{-1}$.

(iv) If $[P'(x)]^{-1}$ exists, then $\Gamma(x)$ is uniquely defined and $\Gamma(x) = [P'(x)]^{-1}$.

(v) If $\Gamma(x)$ is onto, then $P'(x)$ exists and has the inverse $[P'(x)]^{-1} = \Gamma(x)$.

PROPOSITION 1.1.1. If $P'(x)$ and $\Gamma(x)$ exist, then $P'(x)$ maps X onto Y and there is a projection of X onto the kernel of $P'(x)$, i.e. $X = N \dot{+} X_N$ (direct sum), where $N = \{h: P'(x)h = 0, h \in X\}$ and X_N is the range of $\Gamma(x)$.

Note that this concept of an inverse derivative is considered in the strong (Fréchet) sense. However, it is also possible to introduce the notion of an inverse derivative in the weaker sense of Gateaux. (See §2.)

REMARK 1.1.1. Calling $\Gamma(x)$ a right inverse derivative, we could also introduce in a similar way a left inverse derivative $\Gamma(x)$ by using the following formula in place of (1.1.1):

$$||h||^{-1}||\Gamma(x)Q(x)h-h|| \to 0 \text{ as } h \to 0, \ h \in X.$$

1.2. Iteration procedures with inverse derivatives [6].

Our problem is to find a solution to the operator equation

$$(1.2.1) \qquad Px = 0$$

where $P: X \to Y$, X and Y are Banach spaces. We assume the existence of inverse derivatives $\Gamma(x)$ of P in a neighborhood $S(x_0,r) = \{x: ||x-x_0|| \leq r, x \in X\}$, where x_0 is a given approximate solution to (1.2.1). For solving (1.2.1) we use the following iteration procedure:

$$(1.2.2) \qquad x_{n+1} = x_n - \Gamma(x_n)Px_n, \; n = 0,1,2,\ldots .$$

The following theorem gives sufficient conditions for the convergence of the iteration procedure (1.2.2) to a solution of equation (1.2.1).

THEOREM 1.2.1. Suppose that there exist positive numbers $0 < q < 1$, r, η and B such that the inverse derivative satisfies the uniformity condition:

$$(1.2.3) \qquad ||y||^{-1}||P(x+\Gamma(x)y)-Px-y|| \leq q$$

for $x \in S(x_0,r)$ and $||y|| \leq \eta$. Also, suppose that

$$||\Gamma(x)|| \leq B \text{ for } x \in S(x_0,r),$$

$$||P(x_0)|| \leq \eta,$$

$$B\eta(1-q)^{-1} \leq r,$$

and that

P is closed[#] on $S(x_0,r)$.

Then there exists a solution $x^* \in S(x_0,r)$ and the sequence of x_n defined by (1.2.2) converges toward x^*, i.e.

$$x_n \to x^*, \; Px^* = 0, \; x^* \in S(x_0,r) \text{ and}$$

$$(1.2.4) \qquad ||x_n-x^*|| \leq Bq^n(r-q)^{-1}.$$

[#]The operator P is said to be <u>closed</u> if $x_n \in D$, $x_n \to x$ and $Px_n \to y$ imply $x \in D$ and $Px = y$.

REMARK 1.2.1. If $\Gamma(x)$ is onto, then the iteration procedure (1.2.2) becomes the well-known Newton-Kantorovich method [8][14]

$$x_{n+1} = x_n - [P'(x_n)]^{-1} Px_n. \tag{1.2.5}$$

In this case under the hypotheses of Theorem 1.2.1 the solution x^* is unique in $S(x_0,r)$ if, in addition, condition (1.2.3) is satisfied for all $y \in Y$ such that $||\Gamma(x)y|| \leq 2r$, where $x \in S(x_0,r)$. This results from the following inequality

$$(1-q)||y|| \leq ||P(x+\Gamma(x)y)-Px|| \tag{1.2.6}$$

obtained from (1.2.3). For if x^*, $x^{**} \in S(x_0,r)$ are two solutions, then x^* can be written as $x^{**} = x^*+\Gamma(x^*)y$ and we have $||\Gamma(x^*)y|| = ||x^{**}-x^*|| \leq 2r$ and we can apply the inequality (1.2.6).

Consider now the equation $F(x) = 0$, where $F: X \to R$ (reals) is a nonlinear functional on X. If the gradient $F'(x)$ exists and $F'(x)h \neq 0$, where $h \in X$, then we have $t^{-1}|F(x+t[F'(x)h]^{-1}h)-F(x)-t| \to 0$ as $t \to 0$, where $t \in R$, i.e. according to (1.1.2), $\Gamma(x)t = t[F'(x)h]^{-1}h: R \to X$ is an inverse derivative of F at x, if h is fixed and $F'(x)h \neq 0$.

Consider now the following generalization of Newton's method for nonlinear functionals, which is given in [1].

$$x_{n+1} = x_n-[F'(x_n)h_n]^{-1}F(x_n)h_n; \; x_n, h_n \in X. \tag{1.2.7}$$

REMARK 1.2.2. The generalized Newton method (1.2.7) for nonlinear functionals is also a special case of an iteration procedure (1.2.2) with inverse derivatives.

Although the methods (1.2.5) and (1.2.7) are entirely different, both can be considered as particular cases of the procedure (1.2.2).

DEFINITION 1.2.1. We say that $\Gamma(x_0)$ is a <u>uniform inverse derivative</u> of P at x_0 if $\Gamma(x_0)$ is an inverse derivative and if

$$||y||^{-1}||P(x+\Gamma(x_0)y)-Px-y|| \leq q,$$

for $||y|| \leq \eta$ and for x in some neighborhood of x_0.

Using this notion we consider the following modification of procedure (1.2.2)

(1.2.8) $$x_{n+1} = x_n - \Gamma(x_0)Px_n, \quad n = 0,1,2,\ldots .$$

THEOREM 1.2.2. Suppose that there exist positive numbers $0 < q < 1$, r, η and B such that the uniform inverse derivative $\Gamma(x_0)$ satisfies the condition

(1.2.9) $$||y||^{-1}||P(x+\Gamma(x_0)y)-Px-y|| \leq q$$

for $x \in S(x_0,r)$ and $||y|| \leq \eta$, $||\Gamma(x_0)|| \leq B$ and $||Px_0|| \leq \eta$. Then there exists a solution $x^* \in S(x_0,r)$ and the sequence of x_n determined by (1.2.8) converges toward x^*, i.e. $x_n \to x^*$, $Px^* = 0$, $x^* \in S(x_0,r)$ and the error estimate (1.2.4) holds. If $\Gamma(x_0)$ is onto and (1.2.9) is satisfied for all $y \in Y$ such that $||\Gamma(x_0)y|| \leq 2r$, then the solution x^* is unique in $S(x_0,r)$.

The proof of this theorem is exactly the same

as that of Theorem 1.2.1 and is contained in [6].

1.3. Contractors [6].

If we were to analyze the proof of Theorem 1.2.1, we could see that condition (1.2.3) plays a basic role in our argument. This observation leads to the concept of a contractor. Let $P: X \to Y$ be a nonlinear mapping and let $\Gamma(x): Y \to X$ be a bounded linear operator associated with x.

DEFINITION 1.3.1. We say P has a contractor $\Gamma(x)$ if there is a positive number $q < 1$ such that

$$(1.3.1) \qquad ||P(x+\Gamma(x)y)-Px-y|| \leq q\, ||y||,$$

where $x \in X$ and $y \in Y$ will be specified for the problem.

In applications x and y usually run over spheres with centers $x_0 \in X$ and $0 \in Y$, respectively. For instance, if P has a Fréchet derivative $P'(x)$ satisfying

$$(1.3.1^*) \qquad ||P'(x)\Gamma(x)y-y|| \leq q'||y||, \ 0 < q' < 1,$$

then $\Gamma(x)$ is a contractor and condition (1.3.1) is satisfied with $q = (1+q')/2$ and $||y|| \leq \delta||\Gamma(x)||^{-1}$, where δ is chosen so as to satisfy $||P(x+h)-Px-P'(x)h|| < (1-q')2^{-1}\ ||h||$ for $||h|| \leq \delta$. Then for $h = \Gamma(x)y$ we have

$$||P(x+\Gamma(x)y)-Px-y||\leq\{\ ||P(x+\Gamma(x)y)-P'(x)\Gamma(x)y||$$

$$+\ ||P'(x)\Gamma(x)y-y||\}\leq[(1-q')2^{-1}+q']\ ||y||,$$

and $||y|| \leq \delta\ ||\Gamma(x)||^{-1}$ implies $||h|| = ||\Gamma(x)y|| \leq \delta$.

Obviously, an inverse derivative is a contractor. We say that $P: X \to Y$ has a <u>bounded contractor</u> $\Gamma(x)$ if $||\Gamma(x)|| \leq B$ for all x of a certain region.

Suppose now that P has a contractor $\Gamma(x)$ satisfying condition (1.3.1) for all y of Y. Then it is easily seen that the following inequality can be derived from (1.3.1)

$$(1.3.2)\qquad (1-q)||y|| \leq ||P(x+\Gamma(x)y)-Px|| \text{ for } y \in Y.$$

It follows from (1.3.1) that the contractor $\Gamma(x)$ is a one to one mapping and if P is continuous, then (1.3.2) yields the continuity of $[\Gamma(x)]^{-1}$, i.e. then $\Gamma(x)$ is a homeomorphism of Y onto a closed subspace of X. A contractor $\Gamma(x)$ is called <u>regular</u>, if (1.3.1) is satisfied for all $y \in Y$ and $D(P) = \Gamma(x)(Y)$, where D(P) is the domain of P. We say that x is a <u>regular point</u> of P, if $P^{-1}(Px) = \{x\}$ and $Px_n \to Px$ implies $x_n \to x$.

An example of a contractor not involving differentiability can be given as follows.

Let $Px = x - Fx$, where F is Lipschitz continuous with positive Lipschitz constant $q < 1$. Then the contractor inequality for P is

$$||P(x+\Gamma(x)y)-Px-y||=||Fx-F(x+\Gamma(x)y)-(I-\Gamma(x))y||\leq q||y||$$

if $\Gamma(x) = I$ and $Y = X$. Thus, $\Gamma(x) = I$ (the identity mapping) is a contractor for P.

LEMMA 1.3.1. If a contractor $\Gamma(x)$ exists for $x \in D(P)$ and is regular, then x is a regular point of P. If $\Gamma(x)$ is regular and onto, i.e. $\Gamma(x)(Y) = X$, then X is a regular point of P and P is continuous at x. If $\Gamma(x)$ is regular for every $x \in D(P)$, then P has a continuous inverse mapping P^{-1}. If $\Gamma(x)$ is regular and onto for every $x \in D(P)$, then P is a homeomorphism of X onto $P(X)$.

THEOREM 1.3.1. Theorem 1.2.1 remains true if in it we replace the inverse derivative by a bounded contractor $\Gamma(x)$.

1.4. Various iteration procedures as special cases of the contractor method.

It is shown in [2] that the method of steepest descent developed by Kantorovich [8], the minimum residual method investigated by Krasnoselskii and Krein [11] and other gradient methods (See [7], [10], [12].) are special cases of Newton's method (1.2.7) for nonlinear functionals. Thus, these methods can also be considered as special cases of the contractor method. This is the case from variational point of view, i.e. when we reduce the operator equation to the minimum problem of a non-negative nonlinear functional F, where for instance, $F(x) = ||Px||^2 = 0$ is required. However, we can see that these and other methods can also be considered as contractor methods in the direct sense of §1.3.

Let $A: H \to H$ be a linear self-adjoint and positive definite operator in the real Hilbert space H such that $m(x,x) \leq (Ax,x) \leq M(x,x)$ where $0 < m < M < \infty$. Consider the equation

(1.4.1) $$Ax = b; \; x, b \in H.$$

The operator $Px = Ax-b$ is differentiable in the sense of Fréchet and $P'(x) = A$. It is easy to verify that $||\alpha A-I|| < 1$ if $0 < \alpha < 2/M$. Thus, setting $\Gamma(x) = \alpha I$ in (1.3.1*), we obtain a contractor and the corresponding contractor method will be the method of successive approximation with parameter α:

(1.4.2) $$x_{n+1} = x_n-\alpha(Ax_n-b), \; n = 0,1,2,\dots .$$

If in (1.4.2) we replace α by $\alpha(x) = \frac{(r,Ar)}{(Ar,Ar)}$, where $r = r(x) = Ax-b$, we obtain the minimum residual method investigated by Krasnoselskii and Krein [11]:

$$(1.4.3) \qquad x_{n+1} = x_n - \alpha_n r_n, \quad n = 0,1,2,\ldots ,$$

where $\alpha_n = \alpha(x_n)$ and $r_n = r(x_n)$. The contractor here is $\alpha(x)I$ and the contractor inequality (1.3.1) yields in this case $||A(x+\alpha(x)y)-Ax-y|| \leq (M-m)(M+m)^{-1} ||y||$. This inequality is satisfies for $y = r(x)$, by virtue of the following inequality [12, p. 109]

$$(1.4.4) \qquad ||A(x-\alpha(x)r(x))-b|| \leq (M+m)(M-m)^{-1}||r(x)||.$$

To prove the last inequality let us observe that $||A(x-\alpha(x)r(x))-b||^2 = \min_t ||A(x-tr(x)-b||^2 = $
$= \min_t ||r(x)-tAr(x)||^2$. But for $t = 2(M+m)^{-1}$ we have $||I - \frac{2}{M+m} A|| = (M-m)(M+m)^{-1}$ and, consequently, we obtain (1.4.4).

The method of steepest descent developed by Kantorovich [8] for solving (1.4.1) is defined as follows:

(1.4.5) $\quad x_{n+1} = x_n - \beta_n r_n,$

where $\beta_n = (r_n, r_n)(Ar_n, r_n)^{-1}$. Consider the Hilbert space H_A obtained from H by introducing a new scalar product $[u,v] = (A^{-1}u, v)$, $u, v \in H$. Then it is clear that the steepest descent method (1.4.5) is the minimum residual method (1.4.3) considered in the Hilbert space H_A. Thus, the steepest descent method is a contractor method in the Hilbert space H_A. It is easily seen that A is self-adjoint and positive definite in H_A.

Consider the nonlinear operator equation

(1.4.6) $\quad Px = 0,$

where $P: S(x_0, r) \to H$, P is continuously differentiable in the sense of Fréchet in the sphere $S(x_0, r) \subset H$, and $P'(x)$ satisfies the inequality

(1.4.7) $\quad ||P'(x)y|| \geq B^{-1} ||y||$

for all $x \in S(x_0, r)$, $y \in H$. The following iteration procedure [7] is also a contractor method:

$$(1.4.8) \qquad x_{n+1} = x_n - ||Px_n||^2 ||Q(x_n)||^{-2} 2^{-1} Q(x_n),$$

$$n = 0,1,\ldots ,$$

where $Q(x) = [P'(x)]^* Px$ (* = adjoint). Since P is differentiable, we can expect that $\Gamma(x) = ||Px||^2 ||Q(x)||^{-2} 2^{-1} [P'(x)]^*$ will be a contractor, by virtue of (1.3.1*). Condition (1.4.7) implies that $||\Gamma(x)||$ is bounded for $x \in S(x_0,r)$. The existence of a solution of (1.4.6) as well as the convergence of (1.4.8) to this solution can be obtained from Theorem 1.3.1. It follows from the assumptions made in [7] that Theorem 1.3.1 can be applied. It is not difficult to see that the hypotheses made by Kivistik [12, p. 156] are sufficient for the application of the contractor method to his procedure

$$(1.4.9) \qquad x_{n+1} = x_n - (P'(x_n)Px_n, Px_n) ||P'(x_n)Px_n||^{-2} Px_n,$$

$$n = 0,1,2,\ldots .$$

Note that other procedures similar to (1.4.8) can also be put in the unified scheme of the contractor method.

2. Directional Contractors [5].

2.1. Introduction.

The concept of a contractor, considered in §1 as a tool for solving equations in Banach spaces, yields a method for proving existence theorems, as well as convergence theorems, for a broad class of iterative procedures. To prove only existence theorems it is sufficient to define a weaker kind of a contractor that is a directional contractor. The method used in this paragraph combines Gavurin's method of transfinite induction and the concept of a directional contractor.

2.2. Gateaux inverse derivatives and directional contractors.

DEFINITION 2.2.1. Let $P: D(P) \subset X \to Y$ be a nonlinear operator and let $\Gamma(x): Y \to X$ be a bounded linear operator, where X and Y are Banach spaces, and D(P) is a linear subset of X. Suppose that $\Gamma(x)(Y) \subset D(P)$ and

$$P(x+t\Gamma(x)y) - Px - ty = o(t) \tag{2.2.1}$$

for every $y \in Y$, where $||o(t)||t^{-1} \to 0$ as $t \to 0$. Then $\Gamma(x)$ is called the _Gateaux inverse derivative_ of P at $x \in D(P)$.

Thus, this definition implies that $\Gamma(x)$ is one-to-one, and if P has a nonsingular Gateaux derivative $P'(x)$, then $\Gamma(x) = P'(x)^{-1}$ is an inverse Gateaux derivative.

DEFINITION 2.2.2. Suppose now that there exists a positive number $q < 1$ such that

$$||P(x+t\Gamma(x)y) - Px - ty|| \leq qt||y|| \tag{2.2.2}$$

for $0 \leq t \leq \delta(x,y)$. Then we say that $\Gamma(x)$ is a _directional contractor_ of P at x.

It follows from this definition that $\Gamma(x)y = 0$ implies $y = 0$, i.e. $\Gamma(x)$ is one-to-one, and an inverse Gateaux derivative is obviously a directional contractor.

We say that the nonlinear operator P has a _bounded directional contractor_ $\Gamma(x)$ if (2.2.2) is satisfied and, in addition, $||\Gamma(x)|| \leq B$ for all $x \in D(P)$ and some constant B. It is also assumed that $D(P) \subset X$ is linear and $\Gamma(x)(Y) \subset D(P)$ for all $x \in D(P)$.

THEOREM 2.2.1. A closed nonlinear operator $P: D(P) \subset X \to Y$ which has a bounded directional contractor $\Gamma(x)$ is a mapping onto Y.

2.3. A fixed point theorem.

For operators $P = I - F$, where $Y = X$ and I is the identity mapping of X, it is convenient to have contractors of the form $I + \Gamma(x)$. Then the contractor inequality (2.2.2) becomes

$$(2.3.1) \qquad ||F(x+t(y+\Gamma(x)y))-Fx-t\Gamma(x)y|| \leq qt||y||$$

for $0 \leq t \leq \delta(x,y)$, $x \in D(F)$, $y \in X$.
Thus, Theorem 2.2.1 yields immediately

THEOREM 2.3.1. A closed nonlinear operator $F: D(F) \subset X \to X$ which has a bounded directional contractor satisfying condition (2.3.1) and $||\Gamma(x)|| \leq B$, $x \in D(F)$, has a fixed point x^*, i.e. $x^* = Fx^*$. Moreover, $I - F$ is a mapping onto X.

This theorem generalizes the well-known

Banach fixed point theorem. In fact, if $F: X \to X$ is a contraction with Lipschitz constant $q < 1$, then $I + \Gamma(x)$ with $\Gamma(x) \equiv 0$ is obviously a bounded contractor (See [6].) and this notion is much stronger than a directional contractor. However, since the hypotheses of Theorem 2.3.1 are rather weak, we cannot prove the existence of the inverse mapping of $I - F$.

2.4. An application.

We shall now apply Theorem 2.3.1 to prove an existence theorem for nonlinear evolution equations. Consider the initial value problem

$$\frac{dx}{dt} = F(t,x), \ 0 \leq t \leq T, \ x(0) = \xi, \tag{2.4.1}$$

where $x = x(t)$ is a function defined on the real interval $[0,T]$ with values in the Banach space X, and $F: [0,T] \times X \to X$. Denote by X_T the space of all continuous functions $x = x(t)$ defined on $[0,T]$ with values in X and with the norm $||x||_C = \max[||x(t)||: 0 \leq t \leq T]$. Instead of (2.4.1) we consider the integral equation

$$(2.4.2) \qquad x(t) - \int_o^t F(s,x(s))dx = \xi$$

as an operator equation in X_T and we assume that the integral operator is closed in X_T.

For arbitrary fixed $x \in X$ and $t \in [0,T]$ let $\Gamma(t,x)$: $X \to X$ be a bounded linear operator, strongly continuous with respect to (t,x) in the sense of the operator norm. Suppose that there exist positive numbers K and B such that the inequality

$$(2.4.3) \quad \max_{0\leq t\leq T} ||F(t,\{x(t)+\rho\int_o^t \Gamma(s,x(s))y(s)dx\})$$

$$- F(t,x(t))-\rho\Gamma(t,x(t))y(t)|| \leq K\rho||y||_C$$

is satisfied for arbitrary continuous functions $x = x(t)$, $y = y(t) \in X_T$, $0 \leq \rho \leq \delta(x,y)$, where $||\Gamma(t,x)|| \leq B$ for all $x \in X$ and $t \in [0,T]$. Then we say that $F(t,x)$ has a bounded directional contractor $\{I + \int_o^t \Gamma\}$ of integral type.

THEOREM 2.4.1. Suppose that $F(t,x)$ has a bounded directional contractor satisfying (2.4.3) and

T is such that $TK = q < 1$. Then for arbitrary $\xi \in X$ equation (2.4.2) has a continuous solution $x(t)$.

3. Contractors with Nonlinear Majorant Functions.

3.1. Introduction.

In this paragraph a unified theory is developed for a very large variety of iterative methods including the most important ones. By using contractors with nonlinear majorant functions, not only can existence and convergence theorems be proved, but the character of the convergence itself can also be investigated.

3.2. Nonlinear majorants.

Let $P: D \subset X \to Y$ be a nonlinear operator, X and Y being Banach spaces and let $Q(t) \geq 0$, $t \geq 0$, $Q(0) = 0$, be a non-decreasing continuous function. Suppose that $\Gamma(x): Y \to X$ is a bounded linear operator associated with $x \in X$.

DEFINITION 3.2.1. We say that $\Gamma(x)$ is a

contractor for P with <u>nonlinear majorant</u> Q if for $x \in D$; $y, \bar{y} \in Y$; and $x + \Gamma(x)(y-\bar{y}) \in X$ the following inequality is satisfied

$$(3.2.1) \qquad ||P(x+\Gamma(x)(y-\bar{y}))-Px-(y-\bar{y})|| \leq Q(||y-\bar{y}||),$$

where x,y and $\bar{y}$ will be specified for the problem of solving the nonlinear equation

$$(3.2.2) \qquad Px = 0, \; x \in D \subset X.$$

The contractor inequality (3.2.1) can be replaced by the following

$$(3.2.1^*) \qquad ||P(x+\Gamma(x)z)-Px-z|| \leq Q(||z||)$$

for all $x \in S$ and such z that $x + \Gamma(x)z \in D$.

LEMMA 3.2.1. Let $P: D \subset X \to Y$ be a nonlinear operator differentiable in the sense of Fréchet and let its derivative $P'(x)$ be Lipschitz continuous on some sphere $S = S(x_0,r) \subset D$, i.e. there exists a constant K such that

$$||P'(x) - P'(\bar{x})|| \leq K||x-\bar{x}||,$$

for $x, \bar{x} \in S$. Let $A(x)\colon X \to Y$ be a bounded linear nonsingular operator such that

$$||A(x)^{-1}|| \leq B \text{ and } ||P'(x)-A(x)|| \leq C$$

for $x \in S$ and some constants B, C. Then $\Gamma(x) = A(x)^{-1}$ is a bounded linear contractor for P with quadratic majorant $Q(t) = \frac{1}{2}B^2Kt^2 + CBt$, i.e.,

$$||P(x+\Gamma(x)y)-Px-y|| \leq \frac{1}{2}KB^2||y||^2 + CB||y||$$

for $x \in S$ whenever $x + \Gamma(x)y \in S$.

LEMMA 3.2.2. Let $P\colon D \subset X \to Y$ be a closed nonlinear operator and let $T\colon D \subset X \to Y$ be differentiable in the sense of Fréchet on $S = S(x_0, r) \subset D$. Moreover, suppose that $||T'(x)-T'(\bar{x})|| \leq K||x-\bar{x}||$ and $||(Px-Tx) - (P\bar{x}-T\bar{x})|| \leq C||x-\bar{x}||$ for all $x, \bar{x} \in S$, and $T'(x)$ is nonsingular and $||T'(x)^{-1}|| \leq B$ for all $x \in S$.

Then $\Gamma(x) = T'(x)^{-1}$ is a bounded linear contractor for P satisfying the contractor inequality $||P(x+\Gamma(x))-Px-y|| \leq \frac{1}{2}KB^2||y||^2 + CB||y||$, for all

$x \in S$ whenever $x + \Gamma(x)y \in S$.

PROOF. We have $||P(x+\Gamma(x)y)-Px-y||$
$\leq \{||T(x+\Gamma(x)y)-Tx-T'(x)\Gamma(x)y-y||$
$+ ||[P(x+\Gamma(x)y)-T(x+\Gamma(x)y)]-[Px-Tx]||\}\leq\{\frac{1}{2}K||\Gamma(x)y||^2$
$+ C||\Gamma(x)y||\}\leq\{\frac{1}{2}KB^2||y||^2+CB||y||\}$ for all $x \in S$
whenever $x + \Gamma(x)y \in S$. But the contractor $\Gamma(x)$ is said to be bounded on S if there exists a constant B such that

$$||\Gamma(x)|| \leq B \text{ for all } x \in S \subset D. \tag{3.2.3}$$

Q.E.D.

In order to solve (3.2.2) let us consider the following iterative procedures

$$x_{n+1} = x_n - \Gamma(x_n)Px_n, \quad n = 0,1,\ldots \tag{3.2.4}$$

$$t_{n+1} = t_n+Q(t_n-t_{n-1}), \quad t_0 = 0, \quad n = 1,2,\ldots \tag{3.2.5}$$

and assume that the sequence $\{t_n\}$ converges toward some $t^* > 0$. Denote by $S = S(x_0,Bt^*) \subset X$ the open sphere with center x_0, and radius Bt^*.

THEOREM 3.2.1. Suppose that $P: D \subset X \to Y$ is

a closed operator. Let $x_0 \in D$ be chosen so as to satisfy

$$S(x_0, Bt^*) \subset D \text{ and } ||Px_0|| \leq t_1.$$

If P has a bounded contractor $\Gamma(x)$ satisfying (2.2.3) and (3.2.1) then all x_n lie in S and the sequence $\{x_n\}$ defined by (3.2.4) converges toward a solution x of equation (3.2.2) and the error estimate (3.2.6) holds.

$$(3.2.6) \qquad ||x-x_n|| \leq t^* - t_n, \; n = 0,1,\ldots .$$

REMARK 3.2.1. Suppose that the contractor $\Gamma(x)$ has the following property. For arbitrary solutions $x \in S$ and $\overline{x} \in S$ of (3.2.2) there exist y and $\overline{y}$ of Y such that

$$x + \Gamma(x)(y-\overline{y}) = \overline{x}.$$

Then (3.2.2) has a unique solution if $Q(t) < t$ for $t > 0$.

THEOREM 3.2.2. Theorem 3.2.1 remains true with (3.2.1) replaced by (3.2.1*).

REMARK 3.2.2. Putting $Q(t) = qt$ with $0 < q < 1$ in (3.2.1*), we obtain the following contractor inequality

$$||P(x+\Gamma(x)y)-Px-y|| \leq q||y||.$$

In this case $\Gamma(x)$ is a contractor with a linear majorant Q. This is precisely the case investigated in [6].

Now let us consider the case of a contractor with the more general linear majorant $Q(t) = qt + \gamma$, where $0 < q < 1$ and $\gamma \geq 0$. Thus, the contractor inequality is

$$(3.2.7) \qquad ||P(x+\Gamma(x)y-Px-y|| \leq q||y|| + \gamma.$$

THEOREM 3.2.3. Let $P: D \subset X \to Y$ be a closed nonlinear operator. Suppose that $x_0 \in D$ is chosen so as to satisfy $S = S(x_0, Bt^*) \subset D$ and $||Px_0|| \leq \gamma$, where $t^* = \gamma(1-q)^{-1}$. If P has a bounded linear contractor $\Gamma(x)$ satisfying (3.2.3) and the contractor inequality (3.2.7) with $x \in S$ and $||y|| \leq t^*$, then all x_n lie in S and the sequence $\{x_n\}$ defined by (3.2.4) converges toward a solution x of equation (3.2.2) and the error estimate

$$||x-x_n|| \leq \gamma q^n(1-q)^{-1}, \quad n = 0,1,...$$

holds.

3.3. Quadratic majorants.

In the literature there is a variety of iterative methods, which are closely connected with quadratic majorants. (See [14].) These methods and a discussion of them can be included in a general discussion on methods using contractors with quadratic majorants.

Let $P: D \subset X \to Y$ be a closed nonlinear operator which has a contractor $\Gamma(x)$ with quadratic majorant

$$Q(t) = at^2 + bt, \quad a > 0, \; b \geq 0,$$

and which satisfies the contractor inequality

(3.3.1) $$||P(x+\Gamma(x)y)-Px-y|| \leq a||y||^2 + b||y||,$$

whenever $x \in D$ and $x + \Gamma(x)y \in D$. The corresponding majorant iteration procedure is

(3.3.2) $$t_{n+1} = Q(t_n),\ t_0 = \eta,\ n = 1,2,\ldots\ .$$

THEOREM 3.3.1. Suppose that $x_0 \in D$ is chosen so as to satisfy

(3.3.3) $$S = S(x_0, Bt^*) \subset D \text{ and } ||Px_0|| \leq \eta$$

(3.3.4) $$q = a\eta + b < 1,\ t^* = \eta(1-q)^{-1}.$$

If P has a bounded linear contractor $\Gamma(x)$ satisfying (3.2.3) and the contractor inequality (3.3.1) for $x \in S \subset D$ then all x_n lie in S and the sequence $\{x_n\}$ defined by (3.2.4) converges to a solution x of equation (3.2.2) and the error estimate

(3.3.5) $$||x-x_n|| \leq B\eta\, q^n(1-q)^{-1},\ n = 0,1,\ldots$$

holds.

THEOREM 3.3.2. Suppose that $P: D \subset X \to Y$ satisfies the hypotheses of Lemma 3.2.1. If, in addition,

$$||Px_0|| \leq \eta, q = \frac{1}{2}B^2K\eta + BC < 1, r = Bt^*, \text{ and } t^* = \eta(1-q)^{-1},$$

then the sequence $\{x_n\}$ defined by the iteration process $x_{n+1} = x_n - A(x_n)^{-1}Px_n$, $n = 0,1,\ldots$ converges to a solution x of equation (3.2.2); all x_n lie in S; and the error estimate (3.3.5) holds.

By using Lemma 3.2.2 we obtain as a special case of Theorem 3.3.1 the following theorem which is due to Zinčenko [16].

THEOREM 3.3.3. Suppose that in addition to the hypotheses of Lemma 3.2.2 the following conditions are satisfied: $||Px_0|| \leq \eta$, $q = \frac{1}{2}B^2K\eta + BC < 1$, $r = Bt^*$ and $t^* = \eta(1-q)^{-1}$. Then the sequence $\{x_n\}$ defined by the iteration procedure

$$x_{n+1} = x_n - T'(x)^{-1}Px_n, \; n = 0,1,\ldots$$

converges to a solution of equation $Px = 0$. All x_n

lie in S and the error estimate

$$||x-x_n|| \leq B\eta q^n(1-q)^{-1}, \quad n = 0,1,\ldots$$

is valid.

In order to obtain a fixed point theorem we put

$$Px = x-Fx \text{ and } Y = X.$$

Then the contractor inequality (3.2.1) becomes

$$||Fx-F(x+\Gamma(x)(y-\bar{y})) - (I-\Gamma(x))(y-\bar{y})|| \leq Q(||y-\bar{y}||),$$

where I is the identity mapping of X. In the special case where $\Gamma(x) \equiv I$ for all corresponding $x \in D \subset X$, (3.2.1) yields

$$||F(x+(y-\bar{y}))-Fx|| \leq Q(||y-\bar{y}||).$$

Now substituting $\bar{y} = x$ and $y = Fx$ we obtain the contractor inequality

(3.3.6) $$||F^2x-Fx|| \leq Q(||Fx-x||)$$

and this is exactly the case considered by Ortega and Rheinboldt [14, §12.4].

In this case the iteration procedure (3.2.4) becomes the simple successive approximation method

$$x_{n+1} = Fx_n, \quad n = 0,1,\ldots .$$

Thus the successive approximation method with the nonlinear majorant satisfying (3.3.6) is a special case of the general contractor method with the nonlinear majorant satisfying the contractor inequality (3.2.1).

3.4. A special case.

Let us consider the special case of a contractor with quadratic majorant $Q(t) = at^2$, i.e., the contractor inequality is

$$(3.4.1) \qquad ||P(x+\Gamma(x)y)-Px-y|| \leq a||y||^2.$$

THEOREM 3.4.1. Let $P: D \subset X \to Y$ be a closed nonlinear operator with domain D containing the sphere $S = S(x_0,r)$, where $||Px_0|| \leq \eta$, $q = a\eta < 1$, $r = B\eta t^*$ and $t^* = \sum_{n=0}^{\infty} q^{2^n-1}$. Suppose that P has a bounded linear contractor $\Gamma(x)$ satisfying (3.2.3) and the contractor inequality (3.4.1) for $x \in S$ whenever $x + \Gamma(x)y \in S$. Then the sequence $\{x_n\}$ defined by (3.2.4) converges to a solution x of the equation $Px = 0$, all x_n lie in S, and the error estimate

$$(3.4.2) \qquad ||x-x_n|| \leq B\eta \sum_{i=n}^{\infty} q^{2^i-1} \leq B\eta t^* q^{2^n-1}$$

is satisfied.

The following theorem is due to Mysovskih and is also presented in Kantorovitch and Akilov [8]. (See also Ortega and Rheinboldt [14].) This is a special case of Theorem 3.4.1, where $\Gamma(x) = P'(x)^{-1}$.

THEOREM 3.4.2. Suppose that $P: D \subset X \to Y$ is differentiable in the sense of Fréchet in the sphere $S = S(x_0,r) \subset D$, and $P'(x)$ is nonsingular in S and satisfies the following conditions:

$$||P'(x)-P'(\bar{x})|| \leq K||x-\bar{x}|| \text{ for } x,\bar{x} \in S;$$

$$||P'(x)^{-1}|| \leq B \text{ for } x \in S(x_0,r), \text{ where } r = B\eta t^*;$$

$$t^* = \sum_{n=0}^{\infty} q^{2^n-1}; \quad q = \frac{1}{2}B^2K\eta < 1 \text{ and } ||Px_0|| \leq \eta.$$

Then the sequence $\{x_n\}$ defined by Newton's method

$$x_{n+1} = x_n - P'(x_n)^{-1}Px_n, \quad n = 0,1,\ldots$$

converges to a solution x of equation Px = 0. All x_n lie in S and the error estimate (3.4.2) holds.

3.5. Contractors bounded by functions.

So far we have discussed only those cases for which the contractors are bounded by some constants. In this section we consider a more general case.

Let $P: D \subset X \to Y$ be a nonlinear operator and let $x_0 \in D$ be chosen so that $S = S(x_0,r) \subset D$, where r will be defined below. As before let $Q(t), t \geq 0$, be a nondecreasing continuous function with Q(0) = 0. Suppose that $\Gamma(x): Y \to X$ is a linear bounded contractor $\Gamma(x)$ with majorant Q and that the following inequality is satisfied

$$(3.5.1) \qquad ||P(x+\Gamma(x)y)-Px-y|| \le Q(||\Gamma(x)y||)$$

for $x \in S$ and $y \in Y$ whenever $x + \Gamma(x)y \in D$. Suppose further that the following estimate holds for the contractor

$$(3.5.2) \qquad ||\Gamma(x)|| \le B(||x-x_0||)$$

for $x \in S$, where $B(t)$, $t \ge 0$, is a nondecreasing function. Let us consider the sequence $\{t_n\}$ defined by $t_0 = 0$.

$$(3.5.3) \qquad t_{n+1} = t_n + B(t_n)Q(t_n-t_{n-1}).$$

We assume that $t_n \to t^*$ as $n \to \infty$.

THEOREM 3.5.1. Suppose that $P: D \subset X \to Y$ is a closed operator and that $x_0 \in D$ is such that

$$(3.5.4) \qquad ||\Gamma(x_0)Px_0|| \le t_1 \text{ and } r = t^*.$$

If P has a contractor $\Gamma(x)$ satisfying (3.5.1) and (3.5.2), where Q is continuous and nondecreasing with $Q(0) = 0$. Then all x_n lie in S and the sequence $\{x_n\}$ defined by (3.2.4) converges toward a solution x of equation (3.2.2) and the error estimate (3.2.6) holds.

LEMMA 3.5.1. Let $u(t)$ be a real valued function continuous on $[0,t^*]$, where t^* is the smallest root of $u(t) = 0$. Let $v(t) > 0$ for $0 \leq t < t^*$ be continuous on $[0,t^*]$. Suppose that

$$u(s) - u(t) + v(t)(s-t) > 0$$

for arbitrary $0 \leq t < s \leq t^*$.

Then all $t_n < t^*$ and the sequence $\{t_n\}$ defined by

$$(3.5.5) \qquad t_0 = 0, \; t_{n+1} = t_n + u(t_n)/v(t_n), \; n = 0,1,\ldots$$

with initial values $t_0 = 0$, $t_1 = \eta < t^*$ converges toward t^* and satisfies the difference equations

$$(3.5.6) \qquad t_{n+1} - t_n = [1/v(t_n)][u(t_n) - u(t_{n-1}) + v(t_{n-1})(t_n - t_{n-1})].$$

LEMMA 3.5.2. Let $u(t)$ be a real valued function with continuous second derivative $u''(t) > 0$ for $0 \leq t \leq t^*$, where t^* is the smallest root of $u(t) = 0$.

Let $v(t) > 0$ for $0 \leq t < t^*$ be continuous on $[0,t^*]$. If $|u'(t)| \leq v(t)$ for $0 \leq t \leq t^*$, then all $t_n < t^*$ and the sequence $\{t_n\}$ defined by (3.5.5) converges toward t^* and satisfies (3.5.6).

THEOREM 3.5.2. Suppose that $P: D \subset X \to Y$ is a closed operator and $S = S(x_0,r) \subset D$. Assume that P has a contractor $\Gamma(x)$ satisfying the inequalities

$$||P(x+\Gamma(x)y)-Px-y|| \leq Q(||\Gamma(x)y||, ||x-x_0||)$$

for $x \in S$ whenever $x + \Gamma(x)y \in D$, and

$$||\Gamma(x)|| \leq B(||x-x_0||) \text{ for } x \in S.$$

Let the functions $B(t) \geq 0$, $Q(s,t) \geq 0$ for $s,t \geq 0$ be nondecreasing in each variable, $Q(s,t)$ being continuous.

Finally, assume that there exist functions $u(t)$ and $v(t)$ satisfying the hypotheses of Lemma 3.5.1 or 3.5.2 and such that

$$B(s)Q(s-t,t) \quad [1/v(t)][u(s)-u(t)+v(t)(s-t)]$$

for $0 \leq t < s \leq t^*$, where t^* is the smallest root of

$u(t) = 0$; and

$$r = t^*, \; Q(0,t^*) = 0.$$

Then all $x_n \in S$ and the sequence $\{x_n\}$ defined by (3.2.4) converges toward a solution x of $Px = 0$. The error estimate

$$||x-x_n|| \leq t^*-t_n, \; n = 0,1,\ldots,$$

holds with the sequence $\{t_n\}$ defined by (3.5.5) or (3.5.6) and $||\Gamma(x_0)Px_0|| \leq \eta$.

THEOREM 3.5.3. Suppose that $P: D \subset X \to Y$ is a closed operator and $S = S(x_0,r) \subset D$. Assume that P has a contractor $\Gamma(x)$ satisfying

$$||P(x+\Gamma(x)y)-Px-y|| \leq (1/2)K||\Gamma(x)y||^2$$

and

$$||\Gamma(x)|| \leq B/(1-BK||x-x_0||)$$

for $x, x + \Gamma(x)y \in S$, where r, B, K are positive constants such that

$$||\Gamma(x_0)Px_0|| \leq \eta \text{ and } h = Bk\eta \leq 1/2,$$

and $r = t^* = [1-(1-2h)^{1/2}]/(BK)$. Then all x_n lie in S and the sequence $\{x_n\}$ defined by (3.2.4) converges toward a solution x of $Px = 0$ and the error estimate

$$||x-x_n|| \leq t^*-t_n \leq (BK2^n)^{-1}(2h)^{2^n},$$

is valid for $n = 0,1,\ldots$, where the sequence $\{t_n\}$ is defined by

$$t_{n+1}=t_n-BK(t_n-t_{n-1})^2/2(1-BKt_n),t_0=0,t_1=\eta,$$

$n = 0,1,\ldots$ and converges to the smaller root t^* of equation $(1/2)BKt^2-t+\eta = 0$.

Additional theorems using contractors are proved in [3]-[6]. We remark that the use of contractor theory leads to the proof of many results which have appeared in the literature and which use a variety of methods, including the well-known Newton-Kantorovich theorem. As we have seen, these methods are simply special cases of the contractor method. Further results concerning contractor theory will appear elsewhere.

REFERENCES

[1] M. Altman, Concerning approximate solutions of non-linear functional equations, Bull. Acad. Polon. Sci. Cl. III, 5 (1957) 461-465.

[2] M. Altman, Connection between gradient methods and Newton's method for functionals, Bull. Acad. Polon. Sci. Ser. Sci. Math. Astron. Phys., 9 (1961) 887-890.

[3] M. Altman, Contractors and equations in pseudo-metric spaces, to appear.

[4] M. Altman, Contractors, approximate identities and factorization in Banach algebras, Pacific J. Math., to appear.

[5] M. Altman, Directional contractors and equations in Banach spaces, Studia Math., to appear.

[6] M. Altman, Inverse differentiability, contractors and equations in Banach spaces, ibid., to appear.

[7] M. Altman, On the approximate solutions of operator equations in Hilbert space, Bull. Acad. Polon. Sci. Cl. III, 5 (1957) 461-465.

[8] L. Kantorovich and G. Akilov, Functional Analysis in Normed Spaces, Fizmatgiz, Moscow, 1959. D. E. Brown and A. P. Robertson, transl., Pergamon Press, Oxford, 1964.

[9] T. Kato, Nonlinear evolution equations in Banach spaces, Applications of Nonlinear Partial Differential Equations in Mathematical Physics, Proc. Symp. Appl. Math., Vol. 17, Amer. Math. Soc., Providence, 1965, 50-67.

[10] L. Kivistik, On a class of iteration processes in Hilbert space (Russian), Tartu Riikl. Ul. Toimetised, 129 (1962) 365-381.

[11] M. A. Krasnoselskii and S. G. Krein, Iteration process with minimal residuals (Russian), Mat. Sbornik N.S., 31 (1952) 315-334.

[12] M. A. Krasnoselskii, G. M. Vainikko, P. P. Zabreiko, Ya. B. Rutitzkii and B. Ya. Stetzenko, Approximate Solution of Operator Equations (Russian), "Nauka", Moscow, 1969.

[13] J. M. Ortega, The Newton-Kantorovich theorem, Amer. Math. Monthly, 75 (1968) 658-660.

[14] J. M. Ortega and W. C. Rheinboldt, Iterative Solution of Nonlinear Equations in Several Variables, Academic Press, N. Y., 1970.

[15] W. C. Rheinboldt, A unified convergence theory for a class of iterative processes, SIAM J. Numer. Anal., 5 (1968) 42-63.

[16] D. Zinčenko, Some approximate methods of solving equations with nondifferentiable operators (Ukrainian), Dopovidi Akad. Nauk Ukrain. RSR (1963) 156-161.

Acknowledgement. The author is indebted to G. D. Byrne for reading and improving the redaction of this article.

SUBJECT INDEX